Science, Technology, and the
Issues of the Eighties

Also of Interest

† Available in hardcover and paperback.
* American Association for the Advancement of Science Selected Symposia Series.

Westview Special Studies in Science, Technology, and Public Policy/Society

Science, Technology, and the Issues
of the Eighties: Policy Outlook
edited by Albert H. Teich and Ray Thornton for the
American Association for the Advancement of Science

Recognizing that science and technology have become increasingly relevant to important public policy issues, Congress has mandated the periodic preparation of a "Five Year Outlook for Science and Technology" to help U.S. policymakers anticipate and deal with these issues more effectively. This book, the result of a study conducted by the American Association for the Advancement of Science for the second such "Outlook," identifies and explores domestic and international policy concerns in which science and technology are critical factors. The authors' interdisciplinary, nontechnical approach provides policymakers, students and others interested in science, technology and public affairs with a timely overview of nine areas that are likely to become the world's most pressing concerns during the next several years.

Dr. Teich is manager of science policy studies at AAAS and project manager of AAAS's Five Year Outlook Project. Prior to joining AAAS, he was associate professor and deputy director of the Graduate Program in Science, Technology and Public Policy at George Washington University. He is coauthor of *Research and Development: AAAS Reports V and VI* (1980 and 1981) and editor of *Scientists and Public Affairs* (1974) and *Technology and Man's Future* (3rd ed., 1981). **Mr. Thornton** is president of Arkansas State University. He was chairman of the AAAS Committee on Science, Engineering and Public Policy. He served as attorney general of Arkansas and was elected to three consecutive terms in the U.S. House of Representatives. While in Congress he was a member of the House Committee on the Judiciary and chairman of the Subcommittee on Science, Research and Technology.

American Association for the Advancement of Science: Five Year Outlook Project

ADVISORY COMMITTEE

The Honorable Ray Thornton (chairman),[1] President, Arkansas State University, Jonesboro, Arkansas

Dr. R. Darryl Banks, Senior Scientist, RAND Corporation, Washington, D.C.

Dr. William A. Blanpied, Director, Office of Special Projects, National Science Foundation, Washington, D.C.

Dr. Kenneth E. Boulding, Distinguished Professor of Economics, Emeritus, Institute of Behavioral Science, University of Colorado, Boulder, Colorado

Dr. Stothe P. Kezios, Chairman, Department of Mechanical Engineering, Georgia Institute of Technology, Atlanta, Georgia

Dr. Melvin Kranzberg (ex officio),[2] Callaway Professor of the History of Technology, Georgia Institute of Technology, Atlanta, Georgia

Dr. Patricia McFate, Deputy Chairman, National Endowment for the Humanities, Washington, D.C.

Dr. Pauline Newman, Director, Patent and Licensing Department, FMC Corporation, Philadelphia, Pennsylvania

Mr. Rodney W. Nichols, Executive Vice President, The Rockefeller University, New York, New York

Dr. John F. Sherman, Vice President, Association of American Medical Colleges, Washington, D.C.

Dr. Albert H. Teich (project director), Manager, Science Policy Studies, American Association for the Advancement of Science, Washington, D.C.

[1]Chairman, AAAS Committee on Science, Engineering and Public Policy, January 1981–January 1983.
[2]Chairman, AAAS Committee on Science, Engineering and Public Policy, January 1978–January 1981.

Science, Technology, and the Issues of the Eighties: Policy Outlook

edited by Albert H. Teich and Ray Thornton
for the American Association for
the Advancement of Science

A Report Prepared for the National Science
Foundation in Support of the Second Five Year
Outlook for Science and Technology

Routledge
Taylor & Francis Group

LONDON AND NEW YORK

First published 1982 by Westview Press

Published 2019 by Routledge
52 Vanderbilt Avenue, New York, NY 10017
2 Park Square, Milton Park, Abingdon, Oxon OX14 4RN

Routledge is an imprint of the Taylor & Francis Group, an informa business

Library of Congress Cataloging in Publication Data
Main entry under title:
Science, technology, and the issues of the eighties.
    (Westview special studies in science, technology, and public policy/society)
    Based on papers presented at two workshops convened by the American Association for the Advancement of Science Nov. 11–13, 1980, in St. Michael's, Md. and Dec. 10–12, 1980, in Hilton Head, S.C.
    Includes bibliographical references and index.
    Contents: Applying science and technology to public purposes / Richard A. Rettig–The institutional climate for innovation in industry / William J. Abernathy and Richard S. Rosenbloom–Decision making with modern information and communications technology / Donald J. Hillman–[etc.]
    1. Science and state — United States — Congresses. 2. Technology and state — United States — Congresses. I. Teich, Albert H. II. Thornton, Ray, 1928- . III. National Science Foundation (U.S.) IV. American Association for the Advancement of Science. V. Series.
Q127.U6S328   303.4'83   82-1983
                                            AACR2

ISBN 13: 978-0-367-28677-4 (hbk)
ISBN 13: 978-0-367-30223-8 (pbk)

Contents

PART 1
APPLYING SCIENCE AND TECHNOLOGY TO PUBLIC PURPOSES

Preface

Early in 1980 the National Science Foundation (NSF) asked the American Association for the Advancement of Science (AAAS), through its Committee on Science, Engineering and Public Policy (COSEPP), to provide it with assistance in the preparation of the second "Five Year Outlook for Science and Technology." The Outlook, mandated by Congress in the National Science and Technology Policy, Organization and Priorities Act of 1976 (P.L. 94-282), is intended, in part, to identify and describe national problems in which scientific and technological considerations are of major significance and which warrant special attention by policymakers during the next five years. It is also intended to suggest opportunities for—and constraints on—using scientific and technological capabilities to contribute to the resolution of these problems and to achieve other national goals.

The AAAS Five Year Outlook Project was designed to address these public policy elements of the second Outlook. An ad hoc committee, constituted as a subcommittee of COSEPP, served as an advisory body to the project. A list of its members appears at the front of this volume; the full COSEPP membership list appears in Appendix B.

The first task of the Advisory Committee was to identify the issues to be treated in the project. For this purpose, the committee employed a compilation of potentially relevant issues and issue-clusters developed by the AAAS staff on the basis of issues lists obtained from the Congressional Research Service, the House Committee on Science and Technology, the Office of Technology Assessment, the Office of Science and Technology Policy and the AAAS itself, as well as other sources, such as the *New York Times Index*. The committee reviewed the staff compilation and supplemented it with a number of its own suggestions. It then used a set of criteria developed by the staff to evaluate the issues and select those that would be treated in the project. The criteria included:

- *Importance of the issue to the national interest* (higher priority for issues with greater overall impact on the future of the nation and relation to closely held values of the American people);

- *Time-sensitivity* (not just immediacy, but concern for long-range consequences if no action is taken during the next five years);

- *Involvement of science and technology* (centrality of their role in the issue and associated problems or in possible solutions);

- *Breadth and relation to other issues* (definition in a sufficiently broad manner so that the papers could be reviewed and discussed jointly and so each would help illuminate the others);

- *Adequacy of existing institutions to deal with the issue* (higher priority to issues for which existing institutional arrangements are inadequate);

- *Contentiousness* (preference to issues around which there is substantial confusion or misunderstanding);

- *Novelty* (less priority for issues that have received extensive attention recently, especially in the first Outlook).

The Advisory Committee identified two sets of issues — one set centering primarily on a domestic theme, the other on an international theme. For each of the issues, an individual with a solid command of both the academic literature and the policy environment was commissioned to prepare a paper defining the issue, describing what is known about it, including the best available projections of how the issue is likely to develop over the next several decades, and focusing on the policy implications for U.S. science and technology during the next five years.

In order to provide for a careful peer review of each of the papers, as well as to explore the interrelations among the paper topics and to address the larger context within which the topics are embedded, AAAS convened two workshops, one organized around the domestic theme "Applying Science and Technology to Public Purposes," the other around the international theme "Toward Peaceful Change: Science, Technology and International Security."

The Public Purposes workshop was held 10-12 December 1980 in Hilton Head, South Carolina. Thirty-five people attended and drafts of four of the papers from Part 1 of this volume (Chapters 2-5) were presented. The International Security workshop was held 11-13 November 1980 in St. Michaels, Maryland, and was attended by 34 people. First drafts of five of the papers that appear in Part 2 of this report (Chapters 7-11) were presented at this workshop. Names of workshop participants may be found in Appendix A at the end of this volume.

The participants at each workshop constituted a diverse group of experts in various aspects of science, technology and public policy. Each participant was qualified to serve as a technical reviewer for at least one of the commissioned papers. Participants were selected to represent a range of policy perspectives, disciplines, backgrounds and institutional affiliations. They were drawn from universities, government agencies, industrial firms and nonprofit organizations and included demographers, computer and information scientists, geologists, economists, microbiologists, agronomists and persons from a host of other fields.

Drafts of the papers were sent to the participants in advance of each workshop, and each participant brought to the workshop a written review of the paper closest to his or her area of special expertise. The first part of each workshop was devoted to intensive small-group sessions in which the reviewers met with the authors to discuss their comments and to provide the authors with suggestions for revision. Subsequent workshop sessions, both small group and plenary, were devoted to exploration of interrelationships among the papers and to discussion of broader questions surrounding the paper topics.

To capture the outcome of the workshop deliberations, an additional paper — a "synthesis essay" — was commissioned for each workshop. The task assigned to the authors of these synthesis essays was to attend the workshop, to digest the key elements of the papers presented there, as well as the essence of the discussions that took place, and to prepare a paper that addressed the workshop theme and could serve in lieu of formal workshop proceedings. The synthesis essays were conceived as papers that, ideally, would be viewed by the authors of the workshop papers and by the workshop participants as incorporating and fairly representing their papers and deliberations. At the same time, the synthesis authors were expected to draw upon their own knowledge and expertise to provide overall structure, organization and thematic unity that would go well beyond simple reportage.

Revised drafts of the papers presented at the workshops, plus drafts of the synthesis essays, were sent to participants for an additional round of review several weeks after each workshop. Reviews were also solicited from a variety of other individuals who had not attended the workshops. The final products, as presented in this report, have benefited from the reviews and workshop discussions. Nevertheless, it should be pointed out that they remain individual statements. No attempt was made to force consensus among the workshop participants, and many participants, no doubt, disagree with interpretations and policy positions contained in the papers. Similarly, the Advisory Committee and the AAAS, while seeking to assure that the papers and workshops were soundly based and of the highest quality, present them here in order to call attention to and help illuminate discussion of what they regard as vital issues — not to advocate particular points of view.

Numerous individuals contributed to this project and deserve credit for its accomplishments. We can only begin to list them. First we must note the central role of the Advisory Committee. Its distinguished members were extremely conscientious in fulfilling their responsibilities, contributed innumerable ideas, and provided thoughtful guidance throughout the course of the project. We are all deeply in their debt.

Most evident in this report, of course, are the contributions of the authors—both the authors of the original workshop papers and the authors of the synthesis essays. These capable and dedicated individuals labored under extremely tight deadlines, met them, subjected themselves and the products of their labors to intense scrutiny by groups of their peers and cheerfully maintained both their equilibrium and a commitment to the project throughout. Less evident, perhaps, but no less important in the end, were the efforts of the workshop participants, who gave generously of their time and energy to review successive drafts of the papers, to discuss their reviews with the authors and, in a sometimes difficult intellectual exercise, to search for the broader themes linking the papers to one another. All of these individuals, as well as the numerous outside reviewers who provided mail reviews of the papers, have our sincere gratitude.

Thanks are due also to William A. Blanpied and Alan Leshner of the NSF Office of Special Projects, and to William D. Carey, J. Thomas Ratchford and William G. Wells, Jr., of the AAAS, all of whom had oversight responsibility for the project in one sense or another, for their support and guidance, as well as for helping to provide a strong sense of purpose and a commitment to quality in the overall enterprise.

Among those who, in one phase or another, lent their hands, hearts and minds to the project, and to whom we are grateful, are: Andrew Tolmach, summer intern at AAAS, who helped develop the list of candidate issues and selection criteria and helped define the paper topics; Ann Becker and Vicki Killian of Ann Becker and Associates, who guided the development and implementation of the workshop process; Carrie McKee, who edited the final manuscript; Joellen Fritsche and Marlene Povich, who helped design and typed the report; Ginger Payne Keller, who lent her secretarial and administrative skills to the project at several critical points, and Jeanne E. Remington of Westview Press, who ably handled the transformation of the report into a book. A special note of thanks must go to Jill Pace Weinberg, whose title of project secretary/administrative assistant barely begins to suggest the extent of her contributions.

We feel confident that we speak for all of these contributors and the many others from whose advice and assistance we benefited in expressing the sincere hope that this effort will bear fruit in improving our understanding of and ability to handle public policy issues involving science and technology.

Albert H. Teich
Ray Thornton

Observations:
Racing the Time Constants

William D. Carey

If we have learned anything at all about the uses of science and technology in the postwar years, it is that they have an unmistakable influence on contemporary trends and outcomes. They have helped to make the world smaller, spatially, and larger, numerically. They have multiplied our choices and scaled up our risks. They have put men into space and opened a new arena for warfare. They have illuminated man's beginnings and shaken age-old postulates about his worth and destiny. They have unlocked material abundance and laid new burdens on irreplaceable resources. They have expanded man's potential and dramatized his limits. They have advanced clarity and magnified uncertainty. They have penetrated the deepest reaches of knowledge and held a world hostage on the edge of crisis.

We have no reasons to suppose that science and technology will abate their influences upon trends and outcomes and many reasons to expect that they will continue to shape society's choices and dilemmas. What is unprofitable is to try to outguess the rate of advancing knowledge and the forms and effects of its applications through technology. But it is a very different matter to recognize and array the emergent national and global issues confronting the United States and to explore with care the contributions that science and technology could make in managing such issues. What this report for the National Science Foundation seeks to do is to bring to the fore, for Congress and the concerned public, issues of high policy saliency where time is not on our side and where the involvement of science and technology is large and growing.

Left to themselves and to a business-as-usual system of decision making, science and technology will not extricate us from the trouble that is brewing. How science and technology are deployed, toward which goals and at what rates of effort, all depend under our system upon the behavior and the qual-

William D. Carey is the executive director of the American Association for the Advancement of Science, Washington, D.C.

1

ity of the nation's policy apparatus and, to be sure, on the public consensus that legitimizes decision making. In each of the issue-areas with which this report deals, both time and information are central to deciding how the nation's policies are to be positioned and carried out. When lead time is wasted it cannot be recovered. If information is so shallow that policy routes cannot be laid out with confidence, inaction and confusion take the place of resolution. We cannot look to science and technology to dictate policy routes, but we require a broad and deep scientific and technological base upon which to construct and adapt our policy actions. The measure of that base is not represented by the money spent publicly and privately upon research and development but in the appropriateness and the yield of these investments relative to the agenda of salient problems that we face. It is in this sense that policy for science and technology interbreeds with economic, domestic, national security and foreign policies.

The 11 papers that comprise the AAAS contribution to the Five Year Outlook do not begin to resemble a definitive catalog of the issues that will trouble U.S. scientific, technological and public policy over the next five years. But the papers *do* address representative issues of policy, and collectively they have the striking effect of revealing institutional gaps in our national policy machinery relative to dealing with the time constants that are present in varying degrees in the mix of issues.

It is possible to read the report from cover to cover, swimming through the troubled waters sketched by very competent authors, and finally put it down with a sense of an overwhelming agenda. It is possible also to pick and choose among the issues, arraying some in an eclectic structure of priority, although the AAAS itself has not presumed to rank them. But if the report as a whole has objective validity, it should drive us to ask whether our national policy machinery is up to the job of recognizing and dealing with the strategic choices that will be required to bring science and technology's weight to bear, effectively and in time, on the management of these issues. One's basis of confidence, on this score, is very low. The meanings of an exercise in projecting salient policy issues over a five-year period lie in questioning society's institutional capacities to formulate and manage multisectoral strategies aimed toward modifying or altering future outcomes of near-term issues. For a pluralistic society, the institutions favored by command economies are not available. Straight-line policymaking does not fit our constitutional practice. A middle road, on which an informed political consensus is harnessed to decision making, and is driven by a recognition of time constants, is the evident choice.

It might be objected that such counsels veer toward imagining the impossible. Yet, we are now witnessing something that not long ago would have seemed unimaginable, as the government's budgetary, monetary, regulatory and tax strategies interact with the mechanisms of the market economy in designing and pursuing integrated goals for the nation's political economy. If rationalizing the nation's economic agenda is within our institutional capacity, other problems requiring rationalizing may not be out of reach. The lesson could well be that complex public issues, not excluding those involving the timely and effective uses of science and technology, are more likely to respond to coherent "process" management than to pretentious organizational inventions.

Such process innovations are not self-generating, especially in the case of issues that are not built as close to the ground as the state of the domestic economy. With few exceptions, the issues treated in this report are in differential stages of development and calibration. What they have in common, however, is the perceived time constant, which is one thing in the case of population growth and something else (highly uncertain) in the case of materials and energy resource depletion. The makeup of "process" management is not likely to be uniform in dealing with the horizon of issues that we have treated in these papers. From one issue to another, the process would call for different inputs of public policy, long-term corporate strategy, incentives and disincentives, collaborative R&D and upgraded policy research including the social and economic sciences.

Spheres of responsibility will also look different from issue to issue, and it is not the thesis of this report that all responsibility converges on government and its institutions. *Some* responsibility does, however. The national interest is high across the whole array of issues, including those beyond the reach of the United States alone. Because none of the issues is unaffected by government's actions and failures of action, the very minimum responsibility of government is to organize itself in the best sense of "the national security" to keep lively surveillance over the development of issue areas and to see to it that net assessments are made frequently and assimilated by the Congress and the planning arms of our national security and domestic policy machinery.

The residual concern that arises from a study of this kind is not trivial. It is that the problems are outpacing the quality and intensity of our responses, and by widening margins. The potentials of science and technology are not being pressed, much less strained, to meet the national interest. As the lead times shorten, driven by the time constants, risk and vulnerability increase. In the prophetic phrase of Thomas Wolfe, a wind is rising and the rivers flow.

PART 1

Applying Science and Technology
to Public Purposes

1
Applying Science and Technology to Public Purposes: A Synthesis

Richard A. Rettig

Introduction

We live in an era of continuing scientific advance and technological change.[1] This statement, a truism since perhaps the sixteenth century, usefully orients us to the source of our present concerns. Not one but many scientific and technological revolutions go on around us every day; for example, revolutions in consumer electronics, information and communications technologies and molecular biology—all of which are discussed in the papers in this volume. The scope and rate of change severely strain the capacity of existing economic, social and political institutions to respond. "Whirl is king," as Aristophanes wrote, "having deposed Zeus."

Not surprisingly, the range and complexity of scientific and technological change are such that all contemporary societies, industrialized or developing, grapple with the effects of such change through their national governments. Few, if any, scientific and technological issues are of no concern to political leaders and institutions, and few, if any, political issues of moment lack scientific or technological importance.

Quite obviously, in the next five years, in the five years after that, and for the foreseeable future, the United States will confront a number of policy issues that derive in large measure from scientific or technological developments. These are policy issues, moreover, because they have penetrated the social, economic and political fabric of our lives. They constitute an amalgam of accepted scientific and technological knowledge, scientific and technological uncertainty and conflicting political and economic values. And they require collective action to manage if not to resolve.

Richard A. Rettig is chairman, Department of Social Sciences, Illinois Institute of Technology, Chicago, Ill.

It might be assumed that nearly four decades after World War II (the historic watershed of government involvement with the scientific and technological establishments of this country) the management of the scientific-technological enterprise would have become second nature to us. The government's annual investment in research and development, for instance, has grown steadily from $1.2 billion in 1950 to $33.5 billion in fiscal 1980. In addition to direct federal expenditures for research and development (R&D), of course, the private sector of the economy invests comparable funds in R&D, an estimated $29.5 billion, for example, in 1980.

Indeed, there was a period in the early 1960s when optimism about the rate and direction of scientific and technological advance, confidence about our national capacity to manage such advance and belief in the benign effects of the fruits of science and technology were at an all-time high. Since then, however, we as a nation have experienced great difficulties in managing the technological enterprise, especially in bringing science and technology to bear on economic productivity, comparative advantage in international trade and innovation. We have also debated, vociferously at first but with increasing sophistication more recently, the undesirable effects of technological advance; indeed, the potential benefits of scientific advance have even been called into question. The optimism of the early 1960s yielded to pessimism in the late 1960s and early 1970s. Today, one hopes, the nation is moving beyond simple assumptions about the scientific-technological system to a more realistic view of its capabilities and limits and to a constructive recommitment to the use of science and technology for the common good.

The following four papers address in different ways certain of these policy issues. They do not do so exhaustively; rather, they focus selectively on critical aspects of these policy issues. The scope of these papers is limited mainly to the domestic United States, but not entirely so. Issues of international trade, the international nature of environmental pollution and the international character of science itself make it impossible to limit our attention exclusively to the problems of the domestic United States. Nor, as Eugene Skolnikoff's paper argues in the second part of this volume, is it desirable to restrict it in that way.

Several major themes serve to organize this introductory overview essay. First, a policy issue of continuing high-level interest for the past decade is that of encouraging scientific and technological innovation. This issue knows no partisan sponsorship, only perhaps that matters of tone and emphasis may differ from one administration to another.

The second major theme is the need to cope on a continuing basis with the effects of the scientific and technological revolutions that are interlaced with our daily lives. Policy concerns begin with the aforementioned need to encourage innovations of this kind; they reach across all efforts to mitigate the adverse social, political and economic effects of such advance.

Closely related to the second theme is the third, namely, the need to secure a livable world by managing the waste or effluent of an industrial society. The distinction between the two is that the latter is not primarily concerned with scientific and technological innovation as the immediate source of the problem of health, safety and environment, but with all the sources of risk in the society.

Finally, an underlying theme of the papers, and the conference where they were discussed, is that scientific and technological advance has posed deep, perhaps unanswerable, challenges to our established political values, institutions and processes. This last theme is perhaps the most disquieting because few have any clear vision about how to restore existing institutions to a satisfactory level of performance. But diagnosis must precede prescription, and recognition of the problem is thus a constructive step forward.

Encouraging Scientific and Technological Innovation

The lagging productivity of the U.S. economy is a policy issue of great contemporary importance.[2] Growth in productivity increased at an average annual rate of 2.4 percent from 1948 to 1973, but from 1973 to 1978 it was estimated to have grown at less than 1 percent.[3] The 1981 report of the Council of Economic Advisers devoted considerable attention to this productivity decline, enumerating as causal factors the effects of government regulation, increases in energy prices, declines in the rate of growth of capital relative to labor and decreases in spending on research and development.[4]

Edward Denison has suggested that the main source of decline is found in a set of determinants called "advances in knowledge."[5] This set includes technological knowledge—of physical properties and how to make, combine and use physical things—as well as managerial knowledge—of business organization and management techniques in the broadest sense. Denison lumps the "advances of knowledge" determinants together with a set of miscellaneous ones (the effects of government taxation and regulation, the rise in energy prices, the shift away from manufacturing to services and changing attitudes to work) in a "residual" category comprising those least easily measurable determinants of productivity. Commenting on what happened since 1973, he says, "It is possible, even probable, that everything went wrong at once among the determinants that affect the residual series." In short, no single factor adequately explains the productivity decline.

In general, there exists a belief that declining R&D investments have contributed to declining productivity and that increased R&D spending has the
potential for contributing to increased productivity. Considerable discussion exists about measuring these relationships, however, as well as about
the appropriate policy instruments for influencing the situation. Some
would respond by increasing federal funds for R&D; others favor changing
the tax laws to encourage R&D expenditures by private firms.

The policy debate about the relations between research and development,
innovation and productivity will occupy an important place on the public
agenda during the next five years. Policy remedies predictably will address
themselves to the aggregate, or macro-economic, level. In this volume, however, the paper by William Abernathy and Richard Rosenbloom stands in
sharp contrast to this aggregate-level policy discussion, focusing in a provocative and instructive way on the issue of managerial knowledge among
the leaders of U.S. industry.

The point of departure for Abernathy and Rosenbloom is the decline of
productivity and innovativeness of U.S. industry. Noteworthy about their
argument is that they are not primarily concerned with factors external to
private firms that might be causing this decline, like inflation, high rates of
taxation and government regulation, factors that would increase the costs
of production for U.S. firms relative to foreign competitors. Rather, they
are troubled by the institutional climate for industrial innovation in U.S.
firms, by the attitudes and practices of U.S. managers.

Abernathy and Rosenbloom argue that the attention of U.S. managers
has been diverted from long-term technological change toward short-term
adaptation to existing product markets. They observe that the analytic detachment that characterizes such managers is rooted in the financial (and
sometimes legal) pathway to corporate leadership, detachment that contrasts with entrepreneurial leadership that derives from "hands on" experience with the R&D, new product, production and marketing activities of the
firm. Contemporary managers, they argue, display a preference for short-
term cost reduction rather than long-term technological investment. These
managers are market oriented but in a particularly narrow way. They rely
heavily upon market research and its ability to reveal consumer preferences,
rather than depend upon the introduction of a new product to tap latent
preferences and upon an educational campaign for altering those preferences.

Corporate growth and diversification, moreover, often result from the
acquisition of companies not closely related to the firm's historic products.
Rather, they are guided by the portfolio theory of financial risk management that results in a corporate strategy of spreading the risk among a number of diverse enterprises in a complex firm.

To illustrate the consequences of these changed attitudes and practices of U.S. managers, Abernathy and Rosenbloom present a case study of the consumer electronics industry. Several features are noteworthy about this case. First, the consumer electronics industry, as Abernathy and Rosenbloom note, is not heavily regulated by the government, neither by traditional forms of economic regulation nor by more recent health, safety and environmental regulation. Thus, the case forces a search for a different explanation of the loss of U.S. market position. Second, this high-technology product area was one in which U.S. firms, two decades ago, held dominant and undisputed leadership; so the case reflects a loss of market position that has resulted from head-to-head competition with the Japanese. Third, Japanese success lay in the ability of Japanese firms to foresee the application of a high-technology field — electronics — to a large consumer product market and a corporate willingness to pursue a strategy to develop that market. That strategy involved a long time-horizon for maturation of results, including a corollary willingness to make and learn from one's mistakes.

The argument is not entirely persuasive. The main hypothesis that the attitudes and practices of U.S corporate managers have contributed to the decline of U.S. international competitiveness is based on the single case of the consumer electronics industry. Thus, it cannot be automatically extended to even the entire electronics industry or to other industrial sectors. The validity of the argument undoubtedly varies across industries. Second, some will regard the case study data as more anecdotal than systematic in character.

Notwithstanding such criticisms, Abernathy and Rosenbloom have challenged the conventional wisdom on the central issue of declining productivity and the loss of the U.S. competitive international position. The important contribution of the paper, and of the Hayes and Abernathy article that preceded it,[6] is that it points to the structural problem of managerial attitudes and practices as being a major factor contributing to the loss by U.S. firms of technological competitiveness. This issue needs to be widely debated and nowhere more vigo₁ ously than in industry itself. The value of the paper has to be weighed as a contribution to the debate about causal mechanisms of and policy implications for the industrial productivity debate.

Scattered evidence exists that such a debate is beginning. A January 1981 story in the *New York Times* notes that "planning with more distant horizons has become a familiar theme among major American businesses, especially those competing in global markets where foreign competitors, particularly Japanese trading companies, have used the technique to achieve big gains in market share."[7] The account describes a shift in several U.S. firms to greater long-term strategic objectives; a grouping of enterprises within firms according to the appropriate, but differentiated, long-term objectives;

and a linking of corporate salaries more directly to the fulfillment of long-term strategic objectives.

The Japanese themselves are participating in this debate. Recently, Akio Morita, who built the Sony Corporation into a worldwide success, pointed to several factors that give Japanese firms an advantage over U.S. firms.[8] These factors include better long-term planning, bonus payments to employees rather than executives and company-oriented rather than skill-centered executive careers.

A different manifestation of the debate, perhaps more directly supportive of Abernathy and Rosenbloom, is a recent *Harvard Business Review* article about the limits of "return-on-investment" (ROI) analytic techniques for the evaluation of the value of research to a corporation.[9] Mechlin and Berg argue that ROI techniques, when applied to research, fail to value research appropriately. Reasons include: (1) a temporal mismatch between the demands for immediate results and the "natural pace of innovation," (2) the unpredictability of results of research and the inability of ROI to assess the value of negative results and (3) the imprecision of measurement when the results of research may benefit many divisions of a company. The problems with ROI are exacerbated in cases where a central research organization exists for the entire corporation: intrafirm technology transfer is not adequately valued; nor is the effective use of slack resources of facilities and personnel; nor the overhead value of a consumer service function or a personnel recruitment agency. The authors urge supplementing ROI analyses with periodic reviews; mixed project selection, some by product managers and some by research laboratory personnel; calculation of ROI of research throughout a product's life cycle; observation of growth in the firm's relevant product line; and analysis of ROI flowing from self-developed products. In short, the article urges sensitivity to the implications of financial analysis for the research investment and corrective actions that shift away from a short-term emphasis to longer time-horizons. The debate appears to have begun.

What are the policy implications of the Abernathy and Rosenbloom argument? First, they address themselves to a problem rooted in the attitudes and practices of corporate managers and thus to an audience of U.S. corporate leaders. Reform of the behaviors they criticize must come from corporate leaders, not from U.S. government officials. Refreshingly, in contrast to the instinctive tendency of many analysts and commentators, they do not turn to the federal government for help.

Second, U.S. firms, in concert with the U.S. government, may need to become more aggressive in seeking long-term international markets. In particular, Abernathy and Rosenbloom implicitly suggest the need for developing greater access to the Japanese domestic markets.

Third, two successive years of double-digit inflation remind us of the present economic context within which managerial behavior occurs. The 1981 report of the Council of Economic Advisers noted that inflation has risen from an underlying rate of about 1 percent in the first half of the 1960s to a present level of 9 or 10 percent. More ominously, the three major episodes of increase have each begun "with a sharp increase in the underlying rate and ended with the rate falling only part way to its original level."[10] So each successive inflationary period has started from a higher underlying rate than its predecessor.

This historical development of the underlying rate of inflation means that a very strenuous effort will be required to significantly decrease the underlying rate. Furthermore, as long as the rate remains high, U.S. managers of whatever stripe will find few incentives to invest heavily in long-term R&D for the purpose of reestablishing technological leadership in particular industries and markets.[11] The need to control inflation is imperative.

Finally, concern for inflation relates closely to legislation to change the tax treatment of R&D. Such legislation, now being considered by Congress, would increase the incentives for investing in R&D with a long time to payoff. Public policies at this level could reinforce reform tendencies within U.S. management in a constructive way.

Two issues are raised beyond immediate policy concerns by the Abernathy and Rosenbloom argument. In recent years, the educational requirement for corporate success has been the Master of Business Administration (M.B.A.) degree. Implicitly, the authors (both professors at the Harvard Business School) are suggesting that the underpinnings of graduate business education need to be reevaluated. This implication deserves further articulation and discussion.

A second avenue of discussion opened by the paper pertains to the diffusion into the public sector of attitudes and practices similar to those deplored by Abernathy and Rosenbloom in the private sector. We may be witnessing a general weakening of commitment by the federal government and federal R&D managers to invest in long-term, high-risk, but potentially high-benefit scientific and engineering research. If this is occurring in parallel to a similar development within the private sector, then the long-term implications for the nation may be quite serious. Is there any evidence to suggest that such a development has been occurring?

Although *evidence* may be too strong a term, there certainly are signs that a long-term shift has been occurring in federal R&D management. Although the Mansfield amendment of 1969, restricting defense research to projects of direct military relevance, was on the statute books for only one year, many believe that it continues in force today.[12] Perhaps it is time to symbolically "repeal" this amendment by asserting that all federal R&D

agencies have a responsibility to invest in R&D that is broadly appropriate to their mission, not just to that which is narrowly pertinent to specific operational capabilities.

The analog in the public sector of the ascendancy of financial and legal professionals to corporate leadership is the growing number of analysts — economists, M.B.A.'s and others — in the federal government. The general effect of such analysts on R&D is toward shortened time-horizons and sharper emphasis on payoff, a bias against long-term R&D investments.

Stronger pressures for payoffs from federal R&D have led in some instances to a misdirected concern for commercialization of R&D results. Better that the government should invest in building the scientific and technological foundations through long-term R&D than that it should try to pick commercial winners.

In university research, scientific equipment is becoming obsolete to an ever-increasing degree.[13] Moreover, academic researchers today devote a substantially larger portion of their time to administrative matters rather than to research. The effects of these trends can only be pernicious over time.

Three general points deserve statement in concluding this section. First, none of the above factors affecting federal R&D, when taken alone, is that consequential. It is the constellation of factors that is significant and the fact that they all move in the same direction. Second, the severity of the problem is not measured on a year-to-year basis, since this year looks very similar to last. But over 5 or 10 years, we see not continuity but discontinuity, a movement away from a commitment to long-term public R&D investments.

Finally, we may be losing sight of the rationale for supporting public R&D. The rationale is that public investment is needed because of an inherent tendency of private firms to underinvest in the generation of the external benefits of R&D, especially at the research and foundation technology end of the spectrum.[14] At a time when private R&D investments are increasingly constrained to short-term payoff projects, it would be quite unwise for public R&D to offer nothing but a mirror image of that phenomenon.

Managing the Effects of Innovation

Although the need to encourage scientific and technological innovation in the U.S. economy is keenly felt, it is also the case that several scientific and technological revolutions are currently going on before our eyes. One area of continuing scientific and technological innovation, now a quarter of a century old, is that of information and communications technology. Another, far less developed at present in its applications, is that of molecular biology, but prospectively no less sweeping in its potential impact on medicine, agriculture and industry.

Donald Hillman, in "Decision-Making with Modern Information and Communications Technology: Opportunities and Constraints," correctly observes that our society is on the threshold of an Information Age. He provides an overview of the technical change occurring in the technologies of information and communications. This change, driven by the continuing evolution in solid-state electronics, is difficult to comprehend because of its rapid rate, because of the merging of the two technologies of information processing and communications and because its manifestations are so widespread and pervasive throughout all aspects of daily life. The office, the factory, the commercial establishment and even the home are being changed by this revolution.

Numerous major policy problems are raised by the impact of the information and communications revolution. The structure of the telecommunications and information industries is being reshaped; the question of individual privacy becomes more pressing; the management of resource data banks requires attention, as does the availability of international telecommunications and information resources. The formulation of public policy under such circumstances is an exceedingly difficult endeavor. Technical changes reducing costs and extending performance impinge forcefully on so many diverse policy areas, testing existing institutional and legal arrangements in each, that it is difficult to imagine coordination at the federal government level by either Congress, the Executive Branch or the independent regulatory agencies. Furthermore, the application of change is so decentralized, so pervasive, that federal policy formulation is complicated by this fact as well. The challenge in policy terms is to deal sequentially and incrementally with each new policy issue in ways that balance a concern for reaping the benefits of technical change, mitigating its adverse effects, and establishing a flexible framework within which the intelligent guidance of change can occur.

Charles Weiner writes about another revolution, the emergence from sustained research in molecular biology of the gene-splicing techniques of recombinant DNA. These techniques are now used to synthesize insulin, interferon and other proteins, such as industrially important enzymes; they also open the possibility of producing nitrogen-fixing feed grains. Strong scientific advance in molecular biology has been underway for several decades; the revolutionary applications of this body of scientific research are only now beginning.

Yet the recombinant DNA research is stamped indelibly in political and scientific minds as a threshold case in the relations of science to society. Whether this will appear true a decade hence, of course, is not clear. But the case represents what Nelkin has described as the renegotiation of the bargain between science and the polity, the bargain being unquestioning public support for science in exchange for a stream of beneficial science-based innovations.[15]

Why recombinant DNA is regarded as a threshold case warrants comment. First, the prospective benefits are potentially so diverse, so significant and, theoretically, so reachable. From medicine to agriculture to industrial processes to environmental quality controls, a range of benefits is within grasp because of the relatively simple, elegant techniques of recombinant DNA.

Second, the potential benefits have been seen since the mid-1970s against a background of apprehension about the risks of recombinant DNA research. Weiner traces the concern for risk from the 1974 request by a group of prominent molecular biologists to their fellow scientists asking them to refrain voluntarily from performing certain experiments, to the 1975 Asilomar conference, to the Cambridge, Massachusetts, City Council debate about local restrictions on research at Harvard and the Massachusetts Institute of Technology, to the Guidelines for Research Involving Recombinant DNA Molecules, promulgated by the National Institutes of Health (NIH).

The NIH guidelines, Weiner notes, have been revised on three successive occasions since first being issued in 1976, each time becoming less restrictive. During the peak of public concern, legislation was proposed to extend the guidelines to industrial laboratories; that legislation was not enacted nor is it likely to be. Among most academic, government and industrial scientists, a consensus exists today that the prospects of risky outcomes from recombinant DNA research have been steadily reduced, if not ruled out entirely. It is this proved reduction in risk, they argue, that justifies the relaxation of NIH guidelines and explains why restrictive legislation was not enacted. As Weiner notes, however, a few scientists and members of the public continue to express a residual concern for risk.

The third reason why the techniques of recombinant DNA are held in awe by so many thoughtful individuals is that they permit laboratory scientists to manipulate the very constituents of human life itself. Whether one believes that man is the product of a long evolutionary process—"thrown up between ice ages by the same forces that rust iron and ripen corn," to use Carl Becker's felicitous phrase—or the foremost expression of divine creation, the prospect of a few sequestered scientists seeking to "improve" the situation is sufficiently breathtaking to give us all pause. This concern, well founded or not, is sufficiently genuine to be in itself a counsel of prudence from society to the scientific community. Many bench scientists dispute the validity of this awe about manipulating the elements of human life, but that it influences public attitudes is incontrovertible.

The fourth, and more immediate, concern raised by recombinant DNA research is the challenge posed by rapid scientific advance to the social institutions of our time. In particular, what threats are posed to the integrity of the university, to open communication among scientists, to a heretofore largely self-regulated scientific community by the revolution in molecular biology?

Several common concerns underlie these two papers by Hillman and Weiner. First, the phenomenon of interest in each case is that of a powerful scientific and technological revolution. One case, the merging of information and communications technologies, represents a maturing effort whose effects are being felt across an incredibly wide array of applications. The case of molecular biology, and mainly the use of the techniques of recombinant DNA, is less developed; but we are on the threshold, in all likelihood, of several decades of far-reaching applications. In a fundamental way, moreover, each case is an instance of science-based technological change. The empirical or craft tradition in technological change will undoubtedly remain important in the years ahead, but the truly revolutionary technological change of the future will very likely be based upon major scientific advance.

Second, we value such scientific and technological change for the power it displays in several different dimensions. Rapidly declining costs characterize solid-state electronics and are likely to typify the applications of recombinant DNA techniques. Greatly increasing capability is a corollary characteristic. Breadth and diversity of application are yet another dimension of change. The power of this technological advance is clear to all.

Third, this valued scientific and technological advance, encouraged by various government policies, must nevertheless be regulated. There no longer exists an easy one-to-one correspondence in belief that automatically equates scientific and technological advance with social, economic and political progress. In the 1980s, our national commitment to scientific and technological advance is tempered by the realization that adverse consequences can result from the applications of such advance. This sober view is not antithetical to science or technology in inspiration, but neither is it uncritically accepting of a belief that science and technology produce unalloyed social beneficence. Perhaps our current situation is more than a mood, more than mere animus; perhaps it bespeaks a deeper understanding of the relations between science, technology and society.

Securing a Livable World

During the 1970s, the regulatory reach of the federal government was greatly extended to new areas of economic and social life, especially for the purpose of reducing health, safety and environmental risks. The signal of wide public support for this development was Earth Day, on 22 April 1970.

The signal of institutional development was the creation of the Environmental Protection Agency in late 1970; other new agencies included the Occupational Safety and Health Administration (and the National Institute for Occupational Safety and Health) and the Consumer Product Safety Commission. Risk assessment emerged as the central analytic enterprise of this new regulatory activity, informing both the development of general policy and decision making about particular cases in dispute.[16]

Several things can be said about this emergence of risk assessment. First, policy formulation and decision making in health, safety and environmental regulation have been permanently altered, and risk assessors have gained a place at the policy table.[17] Second, this alteration in the mix of participants in the policy arena has been facilitated, even required in some instances, by a strong trend to centralize policy control in the federal government. Centralization has had two dimensions: (1) decisions previously made by the private sector are now made jointly by the private and public sectors; and (2) policies once left to the states have now become the responsibility of the federal government. Third, although risk assessors are now policy participants, their assessments do not dominate policy formulation; most, if not all, assessments are not conclusive and reveal unresolved issues of scientific and technical uncertainty; and political decisions about the acceptable and desirable distribution of risks, costs and benefits are required.

In this context, William Lowrance succinctly argues that the problems created for industry and government by risk-reduction regulation "stem as much from problems of societal attitude and decision-making procedure as from deficiencies of technical analysis and performance." He then suggests several heuristic steps to improve risk assessment and increase the likelihood that sound public policy will be articulated.

Rather than evaluate Lowrance's argument here, we can first clarify the social and political context in which it is written by asking several questions: What is the relationship between risk assessment and science and technology? How does risk assessment go beyond the concerns of science and technology? What is the nature of the political problem confronting risk assessment?

There are a number of diverse ties between risk assessment and science and technology. In the first place, certain areas of science are directly concerned with the physical phenomena that constitute the focus of much risk assessment. Broadly speaking, the environmental sciences have been differentiated from the other natural sciences in the past two decades, although the relationship to the earth sciences is often very close. Analyses of ecosystem behavior—a watershed, an air basin, a forest, for example—may be undertaken for scientific or regulatory reasons—or both. And a larger number of scientists in academic institutions are engaged in the environmental sciences today than was ever true before.

Second, certain areas of science have received strong impetus for development from the effort to regulate risks in health, safety and environment. This is apparent, for example, in toxicology, where substantial increases in research have occurred in government, academic and industrial research laboratories as a result of efforts to regulate the toxic effects of chemicals. It may be the case, however, that toxicological research has been devoted more to routine testing than to explicating underlying mechanisms of action.

Third, advances in instrumentation and analytic techniques have vastly extended the ability of man to detect and measure extremely low concentrations of pollutants in air, water and food.[18] These advances in physical measurement have, quite often, reinforced demands for more regulation. Closely related have been technological advances in control technology. Indeed, a new and significant high-technology industry has arisen in response to health, safety and environmental regulation. This industry represents but another tie of risk assessment to science and technology.

Finally, certain industries have been implicated as bearing greater responsibility than others for worker safety and environmental quality. The chemical industry is one of the foremost among these, being the object of risk-assessment concerns for worker safety in handling dangerous materials, consumer product safety for products like asbestos insulation and environmental quality that is affected by direct, widespread introduction of a dangerous chemical substance like PCBs (polychlorinated biphenyls). An industry so dependent upon science in the first instance necessarily requires highly trained, scientific risk assessors.

It is important to observe, however, that risk assessment goes beyond science and technology and embraces a larger set of issues. Risk assessment, and the regulatory regimes in which it is applied, typically concerns the by-products — the effluent — of modern industrial society, whether generated by individuals, private firms or governments, at every stage from extraction to production to distribution and use of the primary products. The immediate products of science and technology may be included in the domain of risk assessment, but the reach of that domain is far larger.

As a result of the scope of the risk-assessment domain, risk assessors are drawn from widely diverse intellectual fields — physical and natural sciences, engineering, operations research, systems analysis, economics, social science, law and medicine. They are affiliated with a range of different institutions — universities, research and analytic institutes (both nonprofit and for-profit), private industrial firms, government regulatory agencies (at all levels of government) and public-interest law firms. Not surprisingly, therefore, competing and conflicting values, preferences and biases inform the assessment effort, regardless of the agreement that may exist on analytic techniques.

In this context, however, analytic techniques can provide powerful assistance to policy formulation. The nature and scope of the particular risk can be clarified, mechanisms of exposure identified, prospective remedies considered and their respective costs and benefits assessed. But two factors limit the utility of risk assessment. First, although many assessments may identify some critical uncertainties associated with the mechanisms or effects of the risk or its remedy, they often are unable to reduce those uncertainties to an insignificant level by analytic or scientific means. And second, conflicts that arise from the inability of risk assessment to develop a comprehensive description of a risky situation and to specify causal relations between insult, effect and remedy can only be resolved by policy officials acting politically—that is, exercising authoritative discretionary judgment about the preferred allocation of risks, costs and benefits for any given situation or class of situations.

Lowrance would improve the process of risk assessment for the purpose of focusing conflict on essential issues and facilitating the political resolution of such conflict. The greater use of comparative analysis, the explicit statement of standards of risk, the specification of risk-management goals, weighing risks in relation to costs and benefits, are all means to this end. He also suggests that we move with alacrity to identify, first, the cases of "negligible risk" and dismiss them, and—at the other pole—the cases of "intolerable risk" and cease creating them. Between these two poles, Lowrance urges that we set priorities for allocating the scarce resources needed to conduct risk assessments.

An interesting contrast is suggested, however, between Lowrance's recommended strategy and Weiner's account of the recombinant DNA "risk-assessment" exercise. The former is rightly concerned about how risk assessment can facilitate the development of political consensus in the resolution of health, safety and environmental controversies. Yet he addresses himself mainly to improving the analytic aspects of risk assessment.

Weiner, on the other hand, reports on a process initiated by concerned scientists that involves a complex dialogue between scientists and the general public. That dialogue has taken place on university campuses, in city councils, at congressional committee hearings and in the NIH Recombinant DNA Research Advisory Committee. The implicit lesson is that concerns for risk have been allayed in large measure because of scientific developments, but also because the process has forced the scientific community into sustained communication with the public.

It is important to juxtapose the analytic and process features of risk assessment. Analysis is essential but cannot provide the "right" answers. And, because value conflict is likely for any specific risk-reduction effort or policy, procedures that encourage the development of political consensus are also essential to the formulation of sound public policy.

The prospects for achieving political consensus about risk assessment in the 1980s, however, may be fragile. The decade of the 1970s, as noted earlier, witnessed great extension of federal regulatory activity in health, safety and environment. By the end of the decade, there was a growing body of opinion that this regulatory impulse had been carried too far and that its excesses needed to be trimmed back. One expectation about the 1980s, therefore, was that rationalization of the new regulatory area might be undertaken, that is, recognition of the merit of the concern for risk, acknowledgment of the excesses of the consequent regulatory burden and a balancing of the competing values in conflict. Subsequent to January 1981, however, and the advent of a new administration, the prospect exists of a far-reaching effort to undo much of the work of the 1970s, not only the excesses and undesirable burdens but the meritorious effects as well. Whether the United States as a society is close to developing a political consensus about the role and purpose of risk assessment, and its attendant regulatory activity, will be revealed in the next few years. That is the essential question, however, since risk assessment is, in the last analysis, a political issue.

The Challenge to Values and Institutions

Throughout the four papers in Part 1 there runs an undercurrent of anxiety about the adequacy of existing societal institutions and processes to deal with the diverse challenges raised by science and technology. This anxiety was even more pronounced in the conference discussions in December 1980 at which the papers were initially presented.

Reasons for this anxiety are suggested by the examples on every hand. Corporate managers of U.S. firms may in many instances be ill suited by training, career plans and orientation to recognize the requirements for maintaining technological leadership. Information and communications technologies are altering the way we work and live and do business, stretching existing legal and institutional frameworks to their limits; yet these scientific and technological revolutions are still guided by increasingly obsolete policies. The commercialization of molecular biology is placing severe strains on many universities, both between faculty members and their institutions and among faculty members themselves. And the demands that give rise to risk assessment — for example, the control of the effects of acid rainfall or the guarantee of safe disposal of radioactive nuclear waste — test society's capacity to devise solutions that are technically, economically and politically satisfactory.

The challenge of science and technology goes to all societal institutions, from private corporations to universities to the legislative and executive councils of the public sector. A principal source of the challenge is the continuing impact of scientific and technological change which, as Skolnikoff puts it, has been central to "the restructuring of nations and of international affairs, particularly in the 35 years since the Second World War." Policy issues have become a complex amalgam of scientific, technological, economic and political factors. Institutional relationships have also become more complex, largely in adaptive response to the impact of scientific and technological change. National and international spheres are more closely related than ever; public and private sector roles are as difficult to define; and society is increasingly organized into large institutions that share responsibility for governance, usually without commensurate authority. Continuing rapid and widespread scientific and technological change, increasing technical complexity and overwhelming institutional complexity — these are the characteristics of the challenge to societal institutions raised by science and technology.

What is the precise nature of the problem posed by science and technology? It consists of three elements. First, the political institutions of popular, democratic control are inadequate to guide the scientific and technological enterprise and mediate its effects. Surprisingly, none of the authors look to the election of public officials as providing expressions of voter preference on policy issues of any scientific or technological consequence. Nor do they look to elected officials in state legislatures or Congress for significant policy guidance, although that might be a reasonable expectation in a democratic society.

The authors, who come from backgrounds in physical science, engineering, applied social science and history, and the scientists, engineers, businessmen and analysts concerned with managing the scientific and technological enterprise, are oriented to the executive agencies of the federal government. Often, however, a great ambivalence exists toward established authority, and great concern exists about the adequate representation of "the public" in policymaking. It is clear that the "notice and comment" means of securing critical information from interested parties, technical experts and the general public draws only tepid support from many in the scientific and technological communities. Nor does reliance upon the court system, for either fact-finding or dispute settlement, engender any deep allegiance from the community of experts.

Nevertheless, the problem remains of how to appropriately consult the public on issues where scientific and technological considerations loom large. Is the public represented by articulate public-interest groups, other parties at interest, or the citizenry at large? Is it to be consulted through the notice-and-comment procedures of administrative rule making, through public hearings, through formal advisory bodies, or how? The irony should not be lost. Having ignored elections and legislatures — the central institutions of public participation in the governance of our society — we then search seriously but with limited success for acceptable substitutes. The dilemma is one of the deeper institutional crises of our times and impinges directly on the issues raised by science and technology.

The second element of the problem is that elite opinion in the United States is deeply divided on the appropriate response to a number of key policy issues affecting the scientific and technological enterprise. Whether the issue is the cause of productivity decline, the effective means of stimulating innovation, the importance of maintaining a strong scientific and technological base or the appropriate strategies for balancing health, safety and environmental concerns against economic considerations, wide divergence of opinion exists among academic, industrial and governmental leaders. The absence of elite consensus contrasts, for example, with the period just after 1957 when the United States responded to the Soviet Union's triumphant initial entry into outer space. Without such consensus, clear signals cannot be given to the scientific and technological enterprise, and popular support for agreed-upon policies cannot be generated.

The final element of our present difficulty is that the scientific and technological establishment itself is left in a vulnerable position and one from which it is unable to exercise strong leadership. The inadequacy of the institutions of popular control to provide guidance to the enterprise is a recent problem and troubling in its own way. But deeply divergent views among elite policy opinion leaders are more unsettling.

Are there responses to this complex challenge posed by the relentless march of science and technology? At one level, it must be remembered, a number of thoughtful men and women grapple with the day-to-day manifestations of the full range of policy issues raised by and affecting science and technology. This daily hand-to-hand combat, so to speak, responds in an important way to the challenges the society faces.

Beyond the attention given to the immediate aspects of the challenge, however, it is necessary that the task of forging political consensus about major national strategies for science and technology receive high priority in all quarters of the interested public. The quality of dialogue that is required to reconcile the complexity of scientific and technological issues with the need for elite consensus and broad public support is high. Deepening and enriching that dialogue should be a matter of concern to all.

Finally, it may be the appropriate historical period to think more seriously about decentralized responses to the challenges of science and technology. Both the revolutions of information and communications technology and of molecular biology are awesome because their current or expected applications penetrate so deeply into so many facets of contemporary life. The capacity of centralized policy formulation by the federal government is taxed perhaps beyond its limits if it attempts to respond to the full scope and complexity of scientific and technological change. The recombinant DNA lesson may represent in an important way, then, an early model of how the scientific and technological communities and the public ought to engage each other as they mutually strive to guide the societal response to science and technology.

These challenges to values and institutions deserve serious, sustained attention by many in the years ahead, because the effects of scientific and technological advance are sufficiently powerful to alter our lives and patterns of social, economic and political organization. If change must come, better that it should be subjected to continuous scrutiny, discussion and debate within the framework of democratic institutions rather than take us unawares.

Notes

1. I use the more neutral terms of *advance* and *change* rather than *progress* because the relation of scientific and technological advance to social progress is a matter of continuing discussion and debate.

2. Frank Press, "Science and Technology in the White House, 1977 to 1980: Part 1," *Science,* vol. 211 (9 January 1981), pp. 142–44.

3. John Walsh, "Is R&D the Key to the Productivity Problem?" *Science,* vol. 211 (13 February 1981), pp. 685–88.

4. *Economic Report of the President* together with *The Annual Report of the Council of Economic Advisers* (Washington, D.C., January 1981), pp. 68–70.

5. Walsh, "Is R&D the Key to the Productivity Problem?"

6. Robert H. Hayes and William J. Abernathy, "Managing Our Way to Economic Decline," *Harvard Business Review*, vol. 58 (July-August 1980), pp. 67–77.

7. Thomas C. Hayes, "Managers Adopting Long-Term Outlook," *New York Times* (11 January 1981), p. 40, in National Economic Survey (Section 12).

8. Henry Brandon, "SONY Head Analyzes Japan's Formidable Edge," *Washington Star* (20 March 1981); and Hobart Rowen, "Management Is a Family Affair, Japan Counsels U.S. Firms," *Washington Star* (12 April 1981).

9. George F. Mechlin and Daniel Berg, "Evaluating Research—ROI Is Not Enough," *Harvard Business Review,* vol. 58 (September-October 1980), pp. 93–99.

10. *Economic Report of the President.*

11. Edwin Mansfield, "Research and Development, Productivity, and Inflation," *Science,* vol. 209 (5 September 1980), pp. 1091–93.

12. Rodney W. Nichols, "Mission-Oriented R&D," *Science,* vol. 172 (2 April 1971), pp. 29–37; see also Section 203, Public Law 91-121, Military Procurement Authorization for 1970, Ninety-first Congress, First Session, 11 August 1969.

13. Bruce L. R. Smith and Joseph J. Karlesky, *The State of Academic Science: The Universities in the Nation's Research Effort* (New York: Change Magazine Press, 1977).

14. Kenneth J. Arrow, "Economic Welfare and the Allocation of Resources for Invention," pp. 609-21 in *The Rate and Direction of Inventive Activity: Economic and Social Factors* (Princeton, N.J.: Princeton University Press, 1962); Richard R. Nelson, Merton J. Peck and Edward D. Kalachek, *Technology, Economic Growth, and Public Policy* (Washington, D.C.: Brookings Institution, 1967), especially pp. 151-210.

15. Dorothy Nelkin, "Threats and Promises: Negotiating the Control of Research," *Daedalus* (Spring 1978), pp. 191-210.

16. William W. Lowrance, *Of Acceptable Risk: Science and the Determination of Safety* (Los Altos, Calif.: William Kaufmann, Inc., 1976).

17. Compare Charles L. Schultze, *The Politics and Economics of Public Spending* (Washington, D.C.: Brookings Institution, 1968).

18. Arthur L. Robinson, "Myriad Ways to Measure Small Particles," *Science,* vol. 212 (10 April 1981), pp. 146-52; and Thomas H. Maugh II, "New Ways to Measure SO_2 Remotely," *Science,* vol. 212 (10 April 1981), pp. 152-53.

2
The Institutional Climate for Innovation in Industry: The Role of Management Attitudes and Practices

William J. Abernathy and Richard S. Rosenbloom

Evidence of decline in the productivity and innovativeness of American industry is being interpreted in various ways as government, industry and academia struggle to comprehend a troubling phenomenon. Some analyses point to such external factors as inflation, taxation and regulation, which have the common effect of increasing costs incurred by U.S. firms in relation to their foreign competitors. Others stress institutional factors, such as industry structure and the attitudes and practices of managers.

The institutional climate for innovation is important to the behavior of industry. By climate we do not mean just a set of factors external to the firm but a set of attitudes and practices observable within business. Certain basic managerial assumptions shared widely within American culture shape competitive strategies. Strategy, in turn, provides the link between firm and environment.[1]

This paper examines a series of innovations in the consumer electronics industry to explore the strategic role of management attitudes and practices in the management of technology.[2] We hope that this case study of an industry will stimulate other empirical examinations of the consequences of prevalent U.S. managerial precepts. The recent history of the consumer electronics industry provides a fertile ground for exploring broader issues. Of particular interest are the contrasts between the strategic behavior of cer-

William J. Abernathy is professor of business administration, Graduate School of Business Administration, Harvard University, Boston, Mass.

Richard S. Rosenbloom is David Sarnoff Professor of Business Administration, Graduate School of Business Administration, Harvard University, Boston, Mass.

tain Japanese competitors in that industry and the behavior of the leading U.S. firms.[3] Analysis of these contrasts can lead to useful insights into fundamental problems facing U.S. industry in the 1980s.

U.S. Management and Economic Decline

Speculation during the Reagan presidential transition about the possible declaration of a national economic emergency dramatized widespread concern about the health of the U.S. economy. Symptoms of fundamental economic difficulty have been emerging for at least a dozen years. These include increasing rates of inflation and unemployment and declining balances of international trade in key industries.

The causes of these basic problems are complex. The relative importance of contributing factors remains a matter of judgment and debate. There is little disagreement, however, that improved utilization of technology can be a vital part of any remedy. The "Stevenson-Wydler Technology Innovation Act of 1980" finds that "industrial and technological innovation in the United States may be lagging when compared to historical patterns and other industrialized nations."[4] Available indicators of national trends in technological innovation, although neither direct nor conclusive, do provide cause for concern. A recent summary of these trends, presented in an article by one of the present authors (Abernathy) with Robert H. Hayes,[5] points out that:

- labor productivity is increasing more slowly in the United States than in most other industrial nations (Table 2.1);
- rates of productivity growth through the U.S. private sector peaked in the mid-1960s (Table 2.2);
- expenditures in industrial research and development (R&D), as measured in constant dollars, also peaked then, in absolute terms as well as in relation to Gross National Product (GNP) (Figures 2.1 and 2.2).

Although some have attributed these trends to economic or political factors, Hayes and Abernathy argue that the central explanation lies in the attitudes and practices of U.S. managers. In their view, success in the world marketplace requires an organizational commitment to compete on technological grounds, by offering superior products or superior manufacturing processes.

As interpreted in "Managing Our Way To Economic Decline," U.S. managers, guided by what they believe are the newest and best techniques for management, have increasingly directed their attention to matters other

Table 2.1. Growth in Labor Productivity Since 1960 (United States and Abroad).

	Average annual percent change	
	Manufacturing 1960-1978	All industries 1960-1976
United States	2.8%	1.7%
United Kingdom	2.9	2.2
Canada	4.0	2.1
Germany	5.4	4.2
France	5.5	4.3
Italy	5.9	4.9
Belgium	6.9*	—
Netherlands	6.9*	—
Sweden	5.2	—
Japan	8.2*	7.5

*1960-1977.

Source: Council on Wage and Price Stability. Report on Productivity (Washington, D.C.: Executive Office of the President, July 1979). Reprinted from Robert H. Hayes and William J. Abernathy, "Managing Our Way to Economic Decline," Harvard Business Review, vol. 58 (July-August 1980), p. 69.

Table 2.2. Growth of Labor Productivity by Sector, 1948-1978.

	Growth of labor productivity (annual average percent)		
Time Sector	1948-65	1965-73	1973-78
Private business	3.2%	2.3%	1.1%
Agriculture, forestry, and fisheries	5.5	5.3	2.9
Mining	4.2	2.0	-4.0
Construction	2.9	-2.2	-1.8
Manufacturing	3.1	2.4	1.7
Durable goods	2.8	1.9	1.2
Nondurable goods	3.4	3.2	2.4
Transportation	3.3	2.9	0.9
Communication	5.5	4.8	7.1
Electric, gas, and sanitary services	6.2	4.0	0.1
Trade	2.7	3.0	0.4
Wholesale	3.1	3.9	0.2
Retail	2.4	2.3	0.8
Finance, insurance, and real estate	1.0	-0.3	1.4
Services	1.5	1.9	0.5
Government enterprises	-0.8	0.9	-0.7

Note: Productivity data for services, construction, finance, insurance, and real estate are unpublished.

Source: Bureau of Labor Statistics. Reprinted from Robert H. Hayes and William J. Abernathy, "Managing Our Way to Economic Decline," Harvard Business Review, vol. 58 (July-August 1980), p. 69.

Figure 2.1. National Expenditures for Performance of R&D as a Percentage of GNP by Country, 1961–1978.*

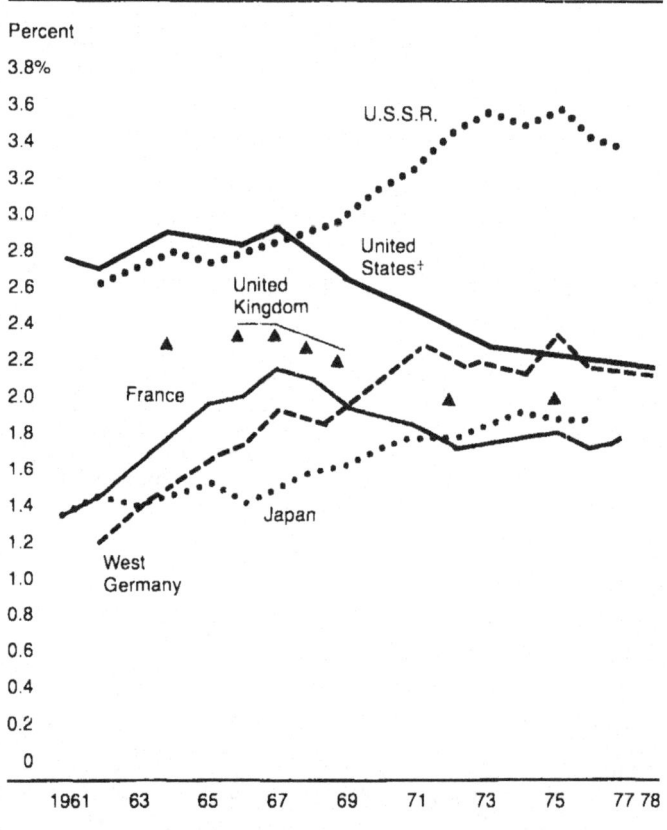

*Gross expenditures for performance of R&D including associated capital expenditures.
†Detailed information on capital expenditures for R&D is not available for the United States. Estimates for the period 1972-1977 show that their inclusion would have an impact of less than one-tenth of 1% for each year.
Source: *Science Indicators – 1978* (Washington. D C.: National Science Foundation. 1979), p. 6.
Note: The latest data may be preliminary or estimates.

Figure 2.2. Industrial R&D Expenditures for Basic Research, Applied Research and Development, 1960-1978 (in Millions of Dollars).

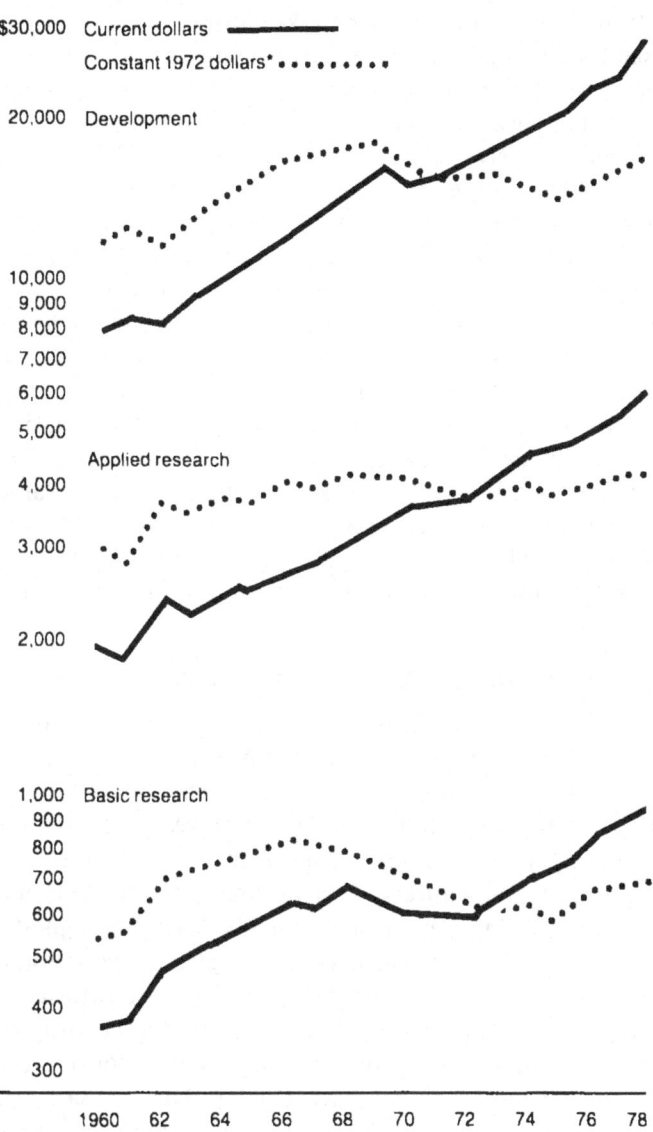

*GNP implicit price deflators used to convert current dollars to constant 1972 dollars.
Source: *Science Indicators – 1978,* p. 87.
Note: Preliminary data are shown for 1977 and estimates for 1978.

than innovation. These techniques, despite their sophistication and wide-spread usefulness, seem to have encouraged analytic detachment at the expense of the insight that comes from "hands on" experience. They promote a preference for short-term cost reduction rather than long-term development of technological competitiveness. According to Hayes and Abernathy, by concentrating on serving existing markets rather than creating new ones, and by excessive emphasis on short-term financial returns and "management by the numbers," many firms seem to have decided against striving for long-term technological superiority as a competitive weapon. They may thus have made themselves vulnerable to competitors whose strategic thrust leads to technological superiority and market leadership.

This provocative thesis attracted widespread attention and acclaim, which the editors of the *Harvard Business Review* believe may have exceeded that of any article ever published in their frequently cited journal. The seven judges for the McKinsey Award, given annually to the best article in the *Review*, unanimously awarded it first prize in 1980.[6]

This thesis is elaborated and made concrete in the following case study of innovation in the consumer electronics industry, which emphasizes certain strategic choices made by Japanese and U.S. firms competing in the U.S. market. While the study suggests certain hypotheses and generalizations, it cannot, of course, "prove" their universality. Although we view this case as an interesting example of the strategic use of technology to gain market leadership, we do not imply that such a strategy is always appropriate.

A Case in Point: Consumer Electronics, 1955–1980

U.S. firms pioneered in consumer electronics technology and until the 1960s took the largest share of revenues and profits in the world's markets.[7] In 1955, U.S. output in consumer electronics was $1.5 billion; Japanese firms produced a mere $70 million. Twenty-five years later the situation is reversed, with Japanese revenues in consumer electronics more than twice those of the U.S. manufacturers. And volume is not the only measure of Japanese leadership. Japanese designs usually offer the highest levels of performance. Unique features, such as the bilingual or stereo sound tracks, are available on television receivers in Japan. Japanese firms also have an overwhelming lead in the most exciting and lucrative new consumer product to reach the electronics industry since the heyday of color television in the mid-1960s: the videocassette recorder (VCR). Factory sales of VCRs, worldwide, exceeded $3 billion in 1980, and 95 percent of those revenues accrued to firms in Japan.

How did this startling reversal come about? More to the point, how did the "imitative" Japanese seize the innovative leadership in this large and important industry? Is this an isolated circumstance or is it the forerunner of more to come? Let us look at the history of the VCR and how a handful of Japanese companies came to dominate a major world market by tracing the history of the consumer electronics industry in the United States and Japan since 1955.

In the mid-1950s, consumer products represented only one-fourth of the output of the U.S. electronics industry; the balance consisted of industrial and military equipment. Yet, in absolute terms, consumer electronics was a big business, accounting for $1.5 billion in factory sales in 1955, the year of peak sales for monochrome television sets. Output of some 8 million television sets, valued at $1 billion, provided two-thirds of the entire consumer electronics industry's revenues that year. (The remainder came largely from radios and phonographs.) But the consumer market for television was by then nearly saturated (88 percent of all homes would have a set by 1960), and demand leveled off in the late 1950s at about 6 million sets per year—98 percent of them black and white. Most sets were either table or console models with screen sizes of about 20 inches measured diagonally. In a highly competitive market, manufacturers standardized their products to gain volume, efficiency and, hence, a lower retail price to the consumer. A few manufacturers tried to introduce "portable" 19-inch models as second sets but without much success. As a result, a shakeout among some 150 television set manufacturers left only 27 in 1960. It would be another five years before the industry experienced another boom with the growth of demand for color television.

The situation in Japan in the 1950s was quite different. Starting from a small base, the output of the Japanese consumer electronics industry increased tenfold from 1955 to 1960. Television set production amounted to only about $30 million in 1955, slightly less than the value of radios produced that year. But by 1960 television sets were already the dominant product in Japan's domestic market—59 percent of the consumer electronics output—and 45 percent of Japanese households had already acquired a television set.

Despite the booming demand for television at home, Japanese firms invested significantly between 1955 and 1960 in opening new markets for radio sets abroad. The story of their 1956 "invasion" of the U.S. radio market with all-transistor portables is well known.[8] Offering a line of miniature receivers half the size and weight of the smallest U.S. products, Japanese producers, led by Sony, developed a major new market segment and met little U.S. competition. Annual sales of portable radios in the United States grew by a factor of seven within a few years, and Japanese imports captured two-thirds of the increase.

By 1960, then, export markets were already significant to the Japanese producers. Products valued at $150 million (87 percent of which were radios) were shipped abroad, representing 20 percent of the Japanese industry's output, while U.S. consumer electronics producers exported only $25 million worth, about 1 percent of their output. As a consequence, by 1960 the scale of consumer electronics production in Japan was already one-third that of the U.S. industry and gaining fast.

The Japanese firms had not yet made inroads into the larger and potentially more profitable television receiver business in the United States, but they were now ready to try. Having achieved an economic scale of production, and possessing a large labor-cost advantage, a Japanese firm might well hope to enter the mainstream of the monochrome television mass market. Some tried in the second half of 1960 with 19-inch sets but without much success; these sets, even at a low price, did not offer sufficient advantages over the established U.S. brands. In 1961 the Japanese share of the U.S. market was a negligible 0.3 percent.

One Japanese firm, rather than competing with the U.S. firms in their areas of strength, chose a different strategy, one reminiscent of its success with the shirt-pocket transistor radio five years earlier. The firm was Sony Corporation, and it introduced a small, lightweight transistorized monochrome receiver with a tiny 8-inch screen at the Chicago Music Show in July 1960, a year before launching it in the U.S. market. Sony was apparently undaunted by the market failure of miniature receivers made in the United States or by U.S. market forecasts. In 1960 the trade press was quoting industry representatives as asserting that significant numbers of U.S. consumers would never buy sets with screens smaller than 17 inches.

But those assertions were wrong. Although transistors were still expensive and the "micro" television sets were a luxury item ($250 when discount houses were selling middle-of-the-line 21-inch sets for under $150), those little sets were unique. In an innovative approach to distribution, Sony established its own sales subsidiary for the United States and sold the microreceivers directly to department stores and other large retailers. Promoted by a highly creative advertising campaign, sales zoomed, rapidly outstripping the still small company's ability to supply the product.

Other Japanese companies soon followed Sony's example, but the U.S. television industry was slow to respond. Not until late 1963 did General Electric (G.E.) become the first company to make a set in this category, an 11-inch transistorized portable. Although most of the other U.S. brands soon filled out their lines with small-screen sets, it was usually with Japanese-made products. The Japanese thus established a base in small-screen sets from which they could expand to larger models and eventually into color.

The numbers tell the story. U.S. imports of Japanese television sets had been negligible in 1960, but the success of micro-television boosted imports to 120,000 units in 1962 and a million in 1965 — one in eight of the monochrome sets sold in the United States. Eighty percent of the Japanese imports had screen sizes of 12 inches or less, representing substantially all of the market for that segment. U.S. companies, now concentrating on the accelerating demand for color receivers, helped boost the demand for these Japanese monochrome sets, for fully 70 percent of the Japanese imports bore U.S. brand names.

Color Television

In the late 1960s the big story in consumer electronics was color television. Color television first took off in the U.S. market, where demand grew from $50 million in 1960 (of $800 million total television receiver sales) to a 1969 level of $2 billion annual factory sales, or 80 percent of all televisions sold. Japan's domestic demand began in 1967 and grew rapidly until 1973. The European color television industry, unable to agree on a technical standard until 1966, did not grow significantly until the early 1970s.

Worldwide factory sales of television receivers (monochrome and color) totaled $6 billion in 1969, with the United States and Japan sharing over three-fourths of those revenues equally. Because color television was a U.S. innovation, pioneered by RCA, one might have expected that the main consequence of its spectacular growth would be to cement the position of U.S. firms as leaders in the world's television manufacturing industry. But the Japanese firms in this industry had already pulled even with the United States in total output — at about $2.3 billion in television sets — and were well on the way to establishing the dominant position they now hold. Japanese firms accomplished this feat in the 1960s by absorbing, then extending foreign technologies; developing a skilled labor force and advanced manufacturing techniques; exploiting their robust domestic market; and adopting export-oriented strategies.

The industry's technological leaders in the 1950s and 1960s had been RCA and N. V. Philips. RCA shared its color television technology under license, as it did all its technical achievements, with firms around the globe. This policy generated significant income over the years for RCA, but it also facilitated foreign entry into the field. N. V. Philips concluded an important agreement with Matsushita in 1952, giving the Japanese firm access to technology in semiconductors, picture tubes and other key components. Matsushita Electronics Corporation, a joint venture between Matsushita and Philips, soon became the largest Japanese producer of these components and a technological leader in its own right. The audiotape cassette, probably the most significant advance in audio magnetic recording, was invented by Philips in the early 1960s and was soon licensed to Sony, Matsushita and other Japanese firms who later became leaders in the consumer audio recording market.

While furthering the technological development of their products, the leading Japanese electronics firms also invested in improved manufacturing processes and in maintaining a highly skilled work force; these steps, too, proved a competitive advantage in the long run.

Japan's large domestic market for consumer electronics, second only to the United States in demand, was also important in the growth of Japanese firms in the 1960s. Behind protective barriers that limited foreign entry, the Japanese firms competed fiercely against each other. New technologies were first introduced on the local market, often as important weapons in their competitive rivalry. Although exports were taken seriously, domestic sales accounted for 60 percent of the revenues of the Japanese consumer electronics producers in 1969.

With exports and domestic sales combined, Japanese producers accounted for nearly two-fifths of the worldwide output of consumer electronics products in 1969, in a country representing only one-quarter of world demand. By then Matsushita, which was the largest Japanese firm in the industry, was the equal of the industry giants – RCA and Philips – with about $1 billion sales in consumer electronics. Moreover, the Japanese were now taking the lead in technology; in 1968 and 1969 two of the most significant technological innovations in color television emerged from Japan: Sony's Trinitron picture tube and Hitachi's all-solid-state color receiver.

With this foundation, the Japanese were prepared in 1970 to take on a new challenge, pioneering in technology for the next major new product in home electronics, the home video player.

Consumer Videocassette Recorders

The videocassette recorder is an innovation of the 1970s. To understand how it came about, however, we must again return to the 1950s. Video recording, like the transistor and color television, was a U.S. innovation. The first practical videotape recorder, which brought important changes in television broadcasting, was introduced by Ampex Corporation in 1956. The machine, called the Quadruplex (or Quad),[9] generated worldwide sales and set the standard for broadcasting use for two decades. Although RCA began producing videotape machines in 1959, Ampex continued to dominate the broadcasting markets and to maintain technological leadership in the Quad family of machines.

In Japan, engineers of more than half a dozen electronics manufacturers called regularly at the studios of NHK, the national television network, to examine the Ampex Quad and pore over its equipment manuals, conferring among themselves and with the NHK engineers. Officials of the Ministry of

International Trade and Investment (MITI) encouraged this interest, giving a small grant to at least one firm to develop videotape technology. Sony, out of its own funds, immediately mobilized a team for the same purpose. Dr. N. Kihara, who had experimented with video recording earlier in the 1950s, headed the team, under the direction of K. Iwama (later president of Sony).

The Quadruplex machine was a massive, complex and expensive machine, filling a large console and two equipment racks. In its monochrome version, it sold for $50,000. The complexity was necessary to produce a signal that met the stringent requirements of broadcast use. As early as 1955, however, Ampex engineers had experimented with an alternative approach, later called helical recording (because early designs wrapped the tape around the rotating cylinder along a helical path). Although the helical recorders also used a rotating head, they were much simpler to make and use than the Quad machines. Early helicals produced pictures that were adequate in quality for the general public but utterly inadequate for the only customers then interested in video recording – the broadcasters.

The Ampex developments in helical technology remained secret until 1962, but a Japanese firm, Toshiba, shocked the industry in 1959 by announcing its patent for a videotape recorder using a helical format. It was clear to engineers that machines built on the helical design could be made smaller and cheaper and would thus be suitable for many uses outside of broadcasting. Not so clear, however, was which path to follow in developing the technology, how to develop new markets for the resulting products and how good a business it would be once products and markets were developed.

During the 1960s, firms in the United States, Japan and Europe participated in the technical and commercial development of helical recording. Outside Japan the leaders were Philips, which dominated European professional and broadcasting markets for video recorders, and Ampex, which extended its broadcast leadership in the United States with a line of professional and industrial units. But neither Philips nor Ampex was focusing on a consumer product at this time. Ampex had always been oriented toward broadcast and professional markets, with consumer products only a minor part of its business, and Philips's consumer divisions were not involved in video recording at all. None of the leading U.S. consumer electronics firms invested significantly in video recording until after 1970.

In Japan, by contrast, eight or more companies – including all the leading consumer electronics manufacturers – launched aggressive efforts to develop helical video recording technology. Sony and Matsushita, among the first to succeed in marketing a consumer product, were consumer electronics companies whose goal, from the very beginning, had been to achieve a design suitable for the home market—even though they sold their first products in other markets. In Matsushita's labs, as early as 1959, an engineer wrote a paper recommending development targeted toward a consumer market, complete with a technical analysis showing the feasibility, in principle, of achieving adequate levels of performance and efficiency. Even earlier, at Sony, Kihara's team first replicated the Quad machine—with knowledge gleaned from the Ampex model but without aid from Ampex engineers—in just a few months. They then set to work on the helical format, guided by a mandate from the company's founder and chief executive, M. Ibuka, who challenged them to build a machine that would cost 1 percent as much as the Quad and could be sold to consumers.

Technical progress, viewed in retrospect, was dramatic, as Table 2.3 indicates. The first Sony product (demonstrated in 1962 and placed on the market in 1963) was one-twentieth the size and one-fifth the price of the Ampex Quad. By 1965 Sony could market a helical machine that used a half-inch tape (versus the more costly two-inch tape of the Quad) and sold at the price that met Ibuka's goal: 1 percent (in constant dollars) of the original Quad machine. Matsushita had a comparable design, and both were shown at the U.S. Consumer Electronics Show in 1965, prompting Ampex to proclaim its intent to market a consumer model the next year.

Despite the fanfare, the videotape recorder of 1965 was still far from being a consumer product. It was still monochrome when the shift to color was already under way. It required manual threading of the tape when experience with the audiocassette had proved that ease of loading made a big difference in consumer demand. And it still used prodigious amounts of very expensive recording tape—even if eight times more efficiently than the Quad format.

Then came the videocassette. By 1970, the first-generation helical cassette machines, developed by Philips in Europe and by several companies in Japan, were ready for demonstration. A worldwide public relations carnival ensued as the press decided that the age of cartridge television in the home had arrived. But they were wrong. Although the technical base was there, it would take until 1975 to develop and market the first successful consumer video player. Meanwhile, in the early 1970s, RCA, Avco, Ampex and others sought to introduce consumer cassette recorders and failed; Sanyo, Toshiba and Matsushita did the same. The only commercially successful products at this stage were destined for professional and industrial use. These were the Philips videocassette recorder, dominant in Europe, and Sony's U-Matic, which set the standard for the now ubiquitous three-quarter-inch "U-format" adopted by Matsushita and Japan Victor.

In the Sony lab, Kihara and his associates took the next logical step beyond the U-Matic and produced the now-legendary Betamax, which was

Table 2.3. Milestones in VTR Product Development.

Market	Model	Company	Date of Commercial Introduction	Tape Width*	Tape Utilization (sq. ft./hour)	Price (in constant 1967 $)
Broadcast	VR-1000	Ampex	1956	2"	747	$ 60,000
Professional	VR-1500	Ampex	1962	2"	375	12,000
Industrial	PV-100	Sony	1962	2"	212	13,000
Industrial/ Professional	EL-3400	Philips	1964	1"	188	3,500
Industrial/ Professional	CV-2000	Sony	1965	1/2"	90	600
Industrial/ Professional	N-1500	Philips	1972	1/2"	70	1,150
Industrial/ Professional	U-Matic	Sony	1972	3/4"	70	1,100
Consumer	Betamax	Sony	1975	1/2"	20	850
Consumer	VHS	Japan Victor	1976	1/2"	16	790
Consumer	VR-2020	Philips	1980	1/2"	6	520

*From 1972 onward, all models used cassettes instead of open reels and all used high-energy tape.

ready for commercialization by early 1974 and on the Japanese market in mid-1975. By eliminating the guard band between recording tracks and exploiting the limits of technology in both heads and tapes, Sony designers reduced the amount of tape needed to record an hour of program by 70 percent. Less tape permitted a smaller, less expensive cassette, which in turn permitted a smaller recorder.

Within two years, engineers at Japan Victor, adopting some of Sony's innovations and adding variations of their own, had perfected an alternative design. Termed VHS (Video Home System), it was adopted by Japan Victor's parent firm, Matsushita, and now shares the bulk of the world market with Sony's Beta format.[10] Matsushita announced the production of their 2-millionth VHS machine in late 1980; Sony's sales of Betamax machines reached 750,000 units in 1980 alone. The sole competitor to these products is an innovative Philips design, called the VR-2020, which is also produced under license by Grundig in Europe. Volume manufacturing operations for the Philips VCRs were starting up in 1980.

The Ingredients of Success

What were the main ingredients of success in the videocassette recorder innovation? Why did the U.S. consumer electronics industry fail totally to establish a manufacturing position in this field? These questions have no simple answers. Any answers proposed at this point are open to challenge. Despite the recency of events and the incompleteness of the record, we offer our interpretation as a working hypothesis.

One key to understanding any innovation is to look at the technology. Whereas Ampex's long dominance of the broadcasting market was won by a single brilliant development produced by a small team in a few years, the home videocassette recorder was developed step by step over 20 years, interactively, by nearly a dozen companies worldwide. Various technical advances had to be combined to produce the necessary features and level of performance for the consumer market – advances in magnetic materials for recording tape and recording heads, and in microelectronic circuitry, coupled with imaginative design of tape formats, tape-handling systems and video circuit design. The engineers at Sony, Japan Victor and Matsushita contributed important inventions, but their Beta and VHS machines also contain many elements invented by Ampex, Philips and Toshiba, whose success in the videocassette recorder business is much more limited.

The successful firms in home video, then, are not distinguished from the rest by inventiveness. No single technical advance unlocked the door to engineering and market success. The successful firms are those whose engineering efforts integrated the technologies for the home videocassette recorder. Several conclusions emerge from a review of their efforts.

First, in view of the large number of Japanese firms competing to develop video recording technology, it is not surprising that the VCR innovators were Japanese. In the early 1960s, substantial development efforts existed at Toshiba, Matsushita, Sony, Japan Victor, Sanyo, Ikegami, Akai, Shibaden and perhaps others. Outside of Japan, only Ampex and Philips appear to have mounted comparable development efforts at that time.

While the Japanese electronics firms were developing video technology, they were also investing in their manufacturing systems, nurturing employee relations, effectively engaging the skills of employees at all levels, introducing innovative manufacturing processes and emphasizing quality and productivity throughout. They did so with a view not only to current requirements but to constant improvement for the future. Such steps enabled the major Japanese companies to develop production capabilities superior to those of most U.S. and European firms.

Furthermore, many of these Japanese firms (and especially the three ultimate leaders) were specialists in consumer products. In contrast, Ampex focused on government and broadcast markets, and Philips was highly diversified. (See Table 2.4.) Development efforts at Sony, Matsushita and Japan Victor began with a consumer product as an ultimate goal. In Sony's case the goal included a clear definition of a target cost, imposing an important economic discipline on development.

Finally, there was the element of persistence. All of the participants in development of the technology tried to commercialize a consumer product prematurely and failed. The Betamax was, in fact, the fourth video recorder generation demonstrated by Sony as a "consumer" product. In 1973, Matsushita geared up an entire department of 1,200 employees to produce a home videocassette recorder that failed utterly in the market. The three current leaders seem to have been able to maintain a strategic commitment that kept development going in the face of disappointment and failure, a strategy similar to that of the more publicized Japanese automotive industry, which had persisted despite initial failures with products first introduced in the U.S. market. They remained committed to small cars, gradually improving product performance, quality and attractiveness to U.S. consumers. These improvements, combined with increased productivity, have given Nissan, Toyota and Honda a quality and production position superior to that of U.S. automotive manufacturers.

The successful innovators in the home videocassette recorder turned out to be, then, consumer electronics companies that had long pursued a global "high-technology" strategy. Their managements foresaw consumer applications of video recording 15 years before the market could actually be tapped, and they persisted in their commitment to develop the basic technology even when prematurely commercialized consumer products failed in

Table 2.4. Strategies of Major Consumer Electronics Producers, 1969.

Company	Total Sales (Million)	Consumer Electronics	Home Appliances	Subtotal
Consumer Electronics Specialists				
Zenith	$ 677	90%	nil	90%
Japan Victor	300	90	nil	90
Sony	300	80*	nil	80*
Consumer Electronics and Appliance Specialists				
Sanyo	$ 500	47%	37%	84%
Matsushita	2,100	50*	25*	75*
Diversified Majors				
Philips	$3,600	28%	10%*	38%*
RCA	3,200	33*	nil	33*
G.E.	8,445	‹5*	N.A.	‹20*
Hitachi	2,300	‹20*	28	‹50*
Toshiba	2,200	20*	20*	41

*Authors' approximations.

Sources: Company annual reports; SEC filings; Japan Company Directory, 1972 (The Oriental Economist).

the market. They had a highly skilled labor force and invested significantly in advanced manufacturing processes. And they were quick to respond to the success of others.

In contrast, the U.S. consumer electronics industry in the late 1950s and throughout the 1960s was held captive by a different ideology. Managements responded to the 1950s market saturation and shakeout by cutting costs. One technical manager said that the standing orders from the television division were to offer them "any new technology available, as long as it gets cost out of the product." Product differentiation was sought in advertising "images" and in such attributes as styling, rather than in performance.

Furthermore, the U.S. industry never developed markets abroad. A senior Zenith executive (later to become the company's chief executive officer), told a Harvard Business School casewriter in early 1972 that "we've always had our hands full with U.S. demand and we've always tended to stick with the biggest payoff and what we knew how to do best. For example, an additional two market share points in the Los Angeles area alone represents more sales volume than there is in most foreign markets." [11]

U.S. managers tended to rely on market research and "objective" analysis to identify latent market opportunities, whereas firms like Sony took risks on novel products and set out to develop the market. For example, in 1955 G.E. had attempted, prematurely as it turned out, to develop the second-set market in the United States with a small-screen monochrome television. In June 1960 (a month before Sony unveiled its micro-television), G.E. management returned to the idea and commissioned market research in which mock-ups of sets of eight different screen sizes, weights and prices were shown to interviewees. The study concluded that "people do not place a high value on portability of the television set." [12]

Throughout the 1960s, while firms like Zenith, G.E. and RCA treated consumer electronics as a mature business with few opportunities for technological leadership, Sony, Matsushita and Japan Victor did the opposite. In radio and then in monochrome and color television, they sought to apply advanced technology to enhance product value to the consumer. Even when domestic demand was brisk, they began to build positions in export markets, starting with the largest (the United States) and aiming at a niche overlooked by the U.S. industry. After initial success, they broadened their lines and deepened penetration.

Their consistent adherence to a high-technology strategy enabled Sony, Matsushita and Japan Victor to take the lead in the videocassette recorder mass consumer market. The technological advantage gained through this strategy was an important ingredient of their success. For example, although U.S. firms designed and built the first miniature all-transistor radios in the 1950s, Sony developed the first product to be successful in the mass market. By capturing the major share of world markets for small radios, Sony and other Japanese producers gained the lead in experience with consumer applications of the transistor. Adapting the transistor to other audio and, later, video products, these firms extended their lead. Similar aggressive strategies in another market segment, audiotape recorders, provided the basis for advantage in another important technology, ferrite recording heads.

Although the largest U.S. firms in the industry, such as G.E. and RCA, also had major technical resources, they were unable to bring them to bear on consumer market opportunities. Some of the barriers limiting their effectiveness in using new technology were organizational. For example, one of the earliest designs for an all-transistor miniature portable radio was developed in the mid-1950s in the corporate lab of one of the giant U.S. companies. The link to the company's own radio business was never forged, and the circuit was licensed to a Japanese producer, who incorporated it into products sold successfully in the United States. Later, the U.S. firm's radio business, in an attempt to catch up, made a photocopy of the Japanese circuit board as the basis for their own design, only to learn that they were copying their own lab's invention. Moreover, the low priority given by the principal U.S. firms to technology for consumer electronics limited their ability to create the technological base necessary to compete in the manufacture of videocassette recorders.

Nor did U.S. firms develop production capabilities competitive in either efficiency or quality; instead they moved their manufacturing facilities to low-cost labor areas like Taiwan or Hong Kong, or they purchased foreign products. They also failed in most cases to develop the unique manufacturing techniques that might have given them a cost, quality or performance edge.

Sony, Matsushita and Japan Victor, unlike most of their American and European competitors, were able to implement their technological strategies through a distinctive organizational system. Technical and commercial staffs at Sony and Matsushita appear to work together effectively, to share information and to understand common goals. Top-level executives, including the chief executive officers, maintained the close contact with the persons developing the new technology that made it possible for their firms to respond rapidly to developments in a constantly changing field.

In contrast, the long-established U.S. firms had to contend with organizational barriers between technical staffs and operating businesses, as illustrated by the transistor radio example above. Their top managers were preoccupied with other priorities and did not involve themselves deeply in VCR technical or marketing issues. Nor did Philips's management; a hint of the significance of these internal factors can be seen in their response to the

Betamax invention. Philips employs a complex and sophisticated organizational structure and resource allocation system to manage its multiple businesses in many countries around the world. While offering many advantages in other respects, that complex organization may have been a real handicap in this innovation. VHS, the Japanese alternative to Betamax, reached the market only 18 months after Sony; it took Philips more than five years to produce their response, the innovative System 2020.

Although the innovative success of Sony, Matsushita and Japan Victor can be attributed primarily to their strategies and organizational methods, they also benefited from their location in Japan. In all of their consumer electronics businesses, they served a large protected domestic market that provided the basic "bread and butter" for cash flow and growth. Furthermore, that market was not fragmented; the leading firms had large shares, giving them a significant scale of operations. The U.S. manufacturers also had a large, concentrated domestic market, but they lacked two things the Japanese had from the start—access to an even larger foreign market and protection against import competition.

The human resource base in Japan may also have provided an advantage. Sony and others could draw on an educated, motivated and stable work force and thus capitalize on skills built up within their companies through the custom of lifetime employment. Because of a unique manufacturing style that integrated effective, if not entirely original, methods of labor management with Japanese cultural traits, the Japanese developed quality and productivity levels that exceeded U.S. capabilities by as much as two to one.

Finally, certain basic cultural factors are evident in Japan's national industrial ideology, which is oriented toward improved quality and efficiency, toward worldwide marketing and toward evaluation of performance on the basis of long-term rather than short-term results. In the context of these underlying management assumptions, the Japanese approach to consumer electronics seems almost inevitable. In terms of the U.S. assumptions, it makes little sense—which brings us back to our main point: at the core of the problem of U.S. competitiveness are the attitudes and practices of U.S. managers.

Managerial Attitudes and Practices

In recent years, the managers of U.S. industry have increasingly preferred to make choices based on abstract analysis of seemingly objective considerations rather than on the insights and judgment of persons seasoned in a business. Concern for near-term financial performance often outweighs long-term considerations. Together, these tendencies produce strategic behavior that is largely reactive. Confronted with effective foreign competitors pursuing proactive strategies, U.S. firms seem to be losing ground.

Not all U.S. firms can be characterized this way, of course. IBM and Texas Instruments, for example, are firms led by managers who combine long-term perspective with deep knowledge of current operations.[13] But these are exceptions to the rule.

It is conventional wisdom that the "push" of new technology yields the greatest rewards when guided by the "pull" of the market. But the paradox of this formulation is that the market's "invisible hand" is expressed through transactions that are possible only after the technology is developed. Hence the rise, in Chandler's apt phrase, of the "visible hand" of the modern corporation to guide the development of technology in anticipation of market rewards.[14]

In many U.S. firms, the methods used to do that job seem to have gotten out of balance, as inappropriate use of common marketing concepts thwarts the incorporation of new technology into innovative products. As illustrated by the contrasting experiences of G.E. and Sony with small-screen television, the needs expressed in the market tend to reinforce the status quo because standard market surveys measure what the customer knows he or she wants now. The initial market estimates for computers, xerographic copiers, the Land camera and other major innovations, for example, fell short by factors of thousands. Successful innovators look beyond expressed needs and lead the market through technologically innovative products that meet latent needs. Formal market analysis is often useful but should not dominate resource allocations to product development.

The very phrase "technology push"[15] may tend to overstate technology's role in the successful introduction of radical innovations. When an innovation captures the market by introducing technologies that address latent needs, significant efforts to educate users about its inherent possibilities have usually been made. The successful videocassette innovators illustrate this point; Sony's brilliant initial advertising of the Betamax as a "time shift" machine is a classic example.

The point is not that product development strategies should always be geared toward latent rather than expressed consumer needs, but that management attitudes and practices geared to the quantifiable and provable, the here and now, risk the loss of such opportunities to use technology to gain competitive advantage.

The conventional wisdom about so-called "mature industries" entails a similar risk of missed opportunities for use of new technology. Mature markets may offer little objective evidence of readiness to accept innovative products, and it is common wisdom for firms competing in them to direct their main efforts to advertising, promoting or reducing the prices of established products. Yet attention to customer needs can reveal opportunities for rewarding investments in technology to differentiate products in performance terms.

The assumption that competitive priorities should change systematically along the life cycle of a product is valid, but should not be followed blindly. The potential value of technological advances in products and processes does not decrease simply because known customer needs have been met. While U.S. managers in the television industry were focusing on volume expansion and cost reduction when growth leveled off, Japanese firms like Sony continued to study latent consumer needs and to introduce major product improvements. Management approaches that operate according to stages in the life cycle create major competitive handicaps if they discourage continuing innovation to meet underlying customer needs.

The biases in management concepts that favor analytic rather than experiential evidence and short-term rather than long-term results are reinforced by parallel tendencies in today's systems of financial control. Three trends have shaped current U.S. practice: (1) increasing diversification of the businesses engaged in by a single firm; (2) consequent decentralization of operations to semiautonomous "profit centers"; reinforced by (3) the emergence of "scientific" theories of corporate finance.

Since the 1950s, a penchant for diversification has led U.S. firms farther away from their core technologies and markets than it has their counterparts in Europe or Japan. Managers in the United States appear to have an inordinate faith in the portfolio law of large numbers, which holds that to amass enough product lines, technologies and businesses is to cushion against the random setbacks of life. This may be true for portfolios of stocks and bonds, where considerable evidence shows that setbacks are random, but businesses are subject to both random setbacks and carefully orchestrated attacks by competitors. Thomas J. Peters, of McKinsey and Company, in discussing 10 well-managed and successful U.S. companies, notes that all are exceptions to this tendency; each, he says, "is a hands-on operator, not a holding company or a conglomerate." Moreover, he argues, "these companies have achieved unusual success by sticking to what each knows best," resisting the temptation to move into new businesses that look attractive but require corporate skills they do not have.[16]

The more general trend toward diversification has reinforced and been reinforced by application of modern theories of financial portfolio management. These principles have increasingly been applied to the creation and management of corporate portfolios, or clusters of companies and product lines assembled through various modes of diversification under a single corporate umbrella. When applied by a remote group of dispassionate experts primarily concerned with finance and control, who lack hands-on experience, the mechanics of portfolio analysis and related resource allocation push managers even further toward an extreme of caution.

The top managers of highly diversified firms necessarily find themselves unable to relate their own experiences to the vital issues of their operating businesses. Since most of these firms use decentralized organizational structures, the manager of each profit center can be held primarily accountable for results. But how does the top manager judge the operating manager's strategic expenditures if they are risky and unlikely to produce near-term results?

Tendencies toward the near term, and toward quantifiable results, produce a situation in which many U.S. managers — especially in mature industries — are reluctant to invest heavily in the development of new manufacturing processes or creative work force policies. By ignoring the competitive advantage of the latter, as in the case of the automotive industry, they adopted homogenous labor relations as dictated by industry unions. This shortsighted action has limited the scope of competitive maneuvers and left the field of work force productivity to foreign competitors. And many U.S. managers assume that essential advances in process technology can be more easily accomplished through equipment purchase than through in-house equipment design and development. This assumption is less widely shared abroad.

Although managers overseas often seek to increase market share through internal development of advanced process technology, even when their suppliers are highly responsive to technological advances, managers in the United States often restrict investments in process development to those items likely to reduce costs in the short run. This diminishes the opportunity for competitive differentiation. Even if companies develop significant new products through aggressive R&D, to the extent that they use established process technology, they reduce their competitors' lead time for introducing similar products. Not only can investing in the development of process technology make products more profitable, when it yields a proprietary process it can serve as a formidable competitive weapon. Indeed, the barrier to entry into videocassette recorder manufacture by U.S. firms is their lack of process know-how. The product technology is open to all; the real secret lies in the Japanese factories.

In sum, we find that certain "modern" strategic concepts, analytic methods and organizational practices discourage the kind of long-term perspective and risk taking necessary to sustain a high level of technological innovation. We may wonder why the negative consequences of these attitudes and practices have become evident only in recent years. A confluence of trends at work over several decades has resulted, we believe, in a significant shift in balance. Paralleling the trends toward corporate diversification, decentralization and use of new concepts of financial management is the growing acceptance of a certain concept of the "successful manager."

There is widespread belief in both the business community and academia in a concept of the professional manager as a "pseudo-professional" — an individual with no special expertise in a particular industry or technology who, nevertheless, can step into and successfully run an unfamiliar company through strict application of financial controls, strategic concepts and market analysis. Although we do not believe that major competitive choices can be made without careful attention to basic marketing and financial issues, we fear that apparently sophisticated analysis of these factors can mask a shallow understanding of customers and a shortsighted view of financial objectives. Moreover, no matter how well these considerations are understood, they are inadequate without a complementary understanding of the technological issues.

It is a rare individual who commands the necessary depth of understanding in each of the major facets of business strategy: markets, finances and technologies. Good organizational design ensures that the operations of the firm are rooted in specialized units able to concentrate on one of these dimensions. But as top management must blend the specialized knowledge, experience and insight of each unit into an integrated, coherent whole in order to make strategic decisions for the entire company, the training and outlook of these integrators at the top of the managerial pyramid are directly relevant. If these individuals are interested in but one or two aspects of the total competitive picture, if their training includes a narrow exposure to the range of functional specialties, they may be unable to implement the necessary integration.

There have been substantial changes over the last two decades in the training and experience that top executives bring to their jobs. Companies have been placing greater value on education and less on experience. The nation's business scools have produced increasing numbers of M.B.A.s armed with knowledge of the latest concepts and faith in their efficacy. No longer does the typical career provide future top executives with intimate hands-on knowledge of the company's technologies, customers and suppliers. Since the mid-1950s, the percentage of new company presidents whose primary interests and expertise lie in the financial and legal areas rather than in production has substantially increased.[17] In addition, many U.S. companies continue to fill new top management posts from outside their ranks. In the opinion of foreign observers, accustomed to long-term careers in the same company or division, "high-level American executives seem to come and go and switch around as if playing a game of musical

William J. Abernathy and Richard S. Rosenbloom

chairs at an Alice in Wonderland tea party."[18] In Japan, by contrast, executives spend a lifetime in one firm where, increasingly, it is the technical man who becomes the chief executive officer.

Trends in management attitudes toward technology and innovation offer another explanation of the emergence in the 1970s of changes in performance. World War II gave great impetus to technology-based innovation and growth in industry. Belief in science and technology was sometimes carried to extremes, as symbolized by lavishly funded corporate research centers established in country-club settings. As this impetus faded, in the 1960s, managers shifted emphasis toward incremental improvements and efficiency gains, a tendency that has been carried to extremes in the 1970s.

Some of these trends have run their course and even been reversed. Since 1976, the expenditure of industry funds (as distinct from government funds) on R&D has risen faster than inflation. Pace-setting companies like G.E. and du Pont are reemphasizing technology and innovation. Significantly, John Welch, G.E.'s new chief executive, and Edward Jefferson, the new chairman at du Pont, are Ph.D. chemists with experience in important innovations within their companies.

Although these are straws in the wind, they suggest that fundamental and widespread changes in prevailing attitudes and practices are possible in the 1980s.

Conclusion

> *The fault, dear Brutus, is not in our stars, But in ourselves.*
> *— Shakespeare, Julius Caesar, act 1, scene 2*

How can the institutional climate in the United States be made more favorable for industrial innovation? If our analysis of the declining competitive position of the United States is valid, fundamental changes in the attitudes and practices of management are needed to reverse — or halt — this decline. We view these changes as a return to values once well established in U.S. industry — the ability to think toward the future, the willingness to innovate and to take bold risks in developing new technologies, new markᵣ and highly productive manufacturing systems. But we also believe that t' changes may require the adoption of new, creative policies toward laboᵣ lations and toward cooperation with government and universities.

Among the attitudes hindering U.S. competitiveness has been the ten dency to neglect product and process technology as a competitive weapon. Senior managers who are inadequately informed about their industry and the nature and interactions of its parts suppliers, workers and customers, or who have little incentive to consider the long-term implications of their own decisions, are more likely to display this tendency. Tight financial controls with short-term emphasis will also bias managers toward choosing the less innovative, less technologically aggressive alternative. Attitudes that preclude creative work force policies are all too common. The character of competition also plays a role. Recent Japanese success in the automobile and consumer electronics industries is partly the result of longstanding technological and market rivalry among several strong firms; others have been quick to match a successful innovator's formula. The key question is, then, how these tendencies can be changed to foster competition, encourage long-term development of basic technologies, stimulate the often risky commercialization of the results of successful technical efforts and maximize work force effectiveness.

Government policies affect industry both directly and indirectly. Policies and programs that have an important impact on industrial innovation — tax structure, monetary policy, regulation, patent policy and aspects of national science policy — have become primary considerations. These broad policies may nurture the scientific and engineering professions and the economy in general, but they fail to provide sufficient conditions for realization of the potential for industrial innovation.

The government may also try to create incentives for long-term research and development, to cushion the risks of innovation and to encourage competition. But the correct approach is as yet unclear, because the linkages between advances in science and technology and such economic outcomes as productivity and innovation are not well understood. Government agencies can reduce the financial risks of investment in advanced technology, as they have in the past, by serving as customers for innovative products. There are other areas of potential influence: the Carter administration's Domestic Policy Review of Industrial Innovation recommended changes in the tax treatment of technology, but these were never endorsed by President Carter. Nor is the impact of the Federal Trade Commission and Justice Department always clear; in many cases their rules may in fact thwart innovation. In the U.S. consumer electronics industry in 1955, for example, the companies involved met most economic tests for a vital industry free of ͻpoly and barriers to entry, yet innovations failed to emerge.

ιiversities can facilitate long-term development of new technologies by continuing to explore ways of structuring relationships with industry. History suggests that industries with healthy links to first-quality academic work are more robust. The U.S. semiconductor industry and German chemical industries are cases in point. In certain frontier areas of engineering — robots or computer-aided design — U.S. industrial firms are already collaborating with university laboratories, by supplying money and assigning technical personnel to work on the programs. Complementary programs in which university personnel have access to state-of-the-art equipment and techniques within industrial laboratories may also prove mutually beneficial. F. Karl Willenbrock, Dean of the School of Engineering and Applied Sciences, Southern Methodist University, has suggested the possible development of engineering analogs to the medical profession's teaching hospitals, where practice and education coexist.[19] In the field of science, interesting new approaches include the 12-year program of biomedical research sponsored by Monsanto at Harvard Medical School and the 10-year program of research on combustion processes sponsored by Exxon at the Massachusetts Institute of Technology. Universities, by their nature, have advantages in continuity of personnel and long-term perspective; industry brings not only resources but vital information about relevant practical needs.

Management faculties might also reflect on some unintended consequences of current methods of management education. Many of the dysfunctional attitudes and practices discussed here are clearly related to what is taught in financial analysis, marketing, planning and related fields. Research designed to clarify the relationships between technological advances and economic outcomes would also be a valuable university contribution.

As essential as cooperation between industry, government and universities may be, opportunities for it are severely limited by the nature and complexity of U.S. business. We believe that the primary agents of change must be industry's top managers themselves. They provide the real leverage, for senior executives make the most significant decisions. If they are well informed, experienced and committed to excellence and innovation, they can effect the changes that will creatively tap this country's human and natural resources and put U.S. industry back into the competitive position it once held throughout the world.

Notes

1. This idea is developed in "Technological Innovation in Firms and Industries: An Assessment of the State of the Art," by Richard S. Rosenbloom, in P. Kelly and M. Kranzberg, eds., *Technological Innovation: A Critical Review of Current Knowledge* (San Francisco: San Francisco Press, 1978).

2. The contributions of several colleagues who helped shape this paper are gratefully acknowledged, including the work of Karen J. Freeze in assisting in its development.

3. The contrasts noted in this example should not be extrapolated to Japanese and U.S. industry generally. There is some evidence to suggest that the particular

Japanese firms examined here are not typical of industry in Japan.

4. Public Law 96-480.

5. Robert H. Hayes and William J. Abernathy, "Managing Our Way to Economic Decline," *Harvard Business Review,* vol. 58 (July-August 1980).

6. The judges included four chief executives of business, two professors (neither from Harvard) and one government official. The editors are unable to recall a previous case where the vote was unanimous.

7. Consumer electronics is a durable goods manufacturing industry whose principal products are television receivers, radios, phonographs and audio and video tape recorders.

8. For an interesting summary and interpretation, see George R. White and Margaret B. W. Graham, "How to Spot a Technological Winner," *Harvard Business Review,* vol. 56 (March-April 1978). The transistor, invented at Bell Laboratories in the United States, was the invention that opened the door to the creation of the semiconductor industry in which U.S. firms have had leading positions. But the application of the transistor to consumer products is a field in which Japanese firms pioneered.

9. The name derives from the use of four magnetic recording heads, located on the edges of a rapidly rotating drum in contact with a tape that moves laterally past the drum.

10. Japan Victor is quick to point out that the development of the VHS was launched well before they became aware of the Betamax and was an independent invention.

11. Zenith Radio Corporation, Harvard Business School case study, 9-674-095, 1974.

12. G.E. subsequently introduced a model with a 19-inch screen at a low retail price. Interestingly, the research had shown the "most preferred" mock-up to be one with a 10-inch screen with 18-pound weight, priced at $259 if transistorized or $129 in a tube version—not unlike Sony's successful design. Source: General Electric Radio and Television Division, Harvard Business School case study, 9-513-082, 1967.

13. An interesting characterization of U.S. best-managed companies is given by Thomas J. Peters in "Putting Excellence into Management," *Business Week* (21 July 1980), pp. 196–205. Peters analyzes the ingredients of successful management in 10 well-run companies, including the 2 mentioned here.

14. Alfred D. Chandler, *The Visible Hand: The Managerial Revolution in American Business* (Cambridge: Harvard University Press, 1977).

15. The shortcomings of "technology-push" strategies in the public sector have also been noted. William J. Abernathy and Balaji S. Chakravarthy argue, in "Government Intervention and Innovation in Industry: A Policy Framework," *Sloan Management Review* (December 1979, pp. 5-17), that government attempts to innovate through technology push alone have usually failed. Complementary efforts at "technology pull" through intervention in products and markets seem to have been associated with most successful cases.

16. Peters, "Putting Excellence into Management."

17. See Exhibit VII in Hayes and Abernathy, "Managing Our Way to Economic Decline."

18. Not all U.S. companies fit this description. Some well-known companies that emphasize promotion from within—for example, IBM, 3M, Texas Instruments or Citicorp—are also well known for risk taking and innovative success.

19. "Engineering—the Neglected Ingredient," remarks at the Science Policy Seminar, George Washington University, 9 December 1980.

3
Decision Making with Modern Information and Communications Technology: Opportunities and Constraints

Donald J. Hillman

The Growth of Information and Communications Technology

The growth of sophisticated information processing systems, accompanied by huge advances in telecommunications capabilities, constitutes an information revolution that raises significant new issues for policymakers. This paper analyzes the historical and recent developments in data processing and telecommunications, the impact these advances are having on society and the associated policy issues that need to be addressed.

Historical Perspective

A significant aspect of the growth of information and communications technologies is that the two technologies are merging. To gain a better understanding of this phenomenon, it may be helpful to examine three categories that make up the information industry, for it is their interaction that forms the most powerful information processing and communications systems.

The functional categories of importance are:

- information and data processing technologies;
- word processing technologies;
- communications technologies.

Donald J. Hillman is director, Center for Information and Computer Science, Lehigh University, Bethlehem, Pa.

Information and Data Processing. Information and data processing technologies are primarily associated with computers. Advances in computer performance have been spectacular, in both the variety and the number of different applications. Advances in miniaturization have reduced computing costs substantially, thereby expanding the number of users and enlarging possible applications. Two basic architectural features of computers are involved in this rapid progress. The first of these is a *hierarchy of memories.* Memories range from slow, high-capacity peripheral devices (such as magnetic tapes and disks) to fast, limited-capacity central memories (such as magnetic cores or semiconductors) to high-speed registers.

The second feature is the *central processing unit,* which contains the arithmetic/logic unit and a control unit. The arithmetic/logic unit manipulates the high-speed registers according to logical operations. The control unit is responsible for examining the programmer's instructions and for controlling the actions of the memories and the arithmetic/logic unit to perform the necessary operations. Instructions and data are both stored in the same memory. In the future are fundamental architectural changes in computer design, which promise to increase performance significantly.

Word Processing Technology. Word processing manipulates text without regard to message or semantic content. Text editing, while closely identified with word processing, is used to align the formats of large multipage documents and reports and to handle routine office correspondence. Word processing was applied initially to manual office functions, and it affected primarily secretarial and clerical workers. There is growing evidence, however, that word processing is beginning to make an impact on office operations in general, and that when linked with data processing, the combination will substantially alter future management styles.

Communications Technology. The telephone has been transformed from a signal transport device into a message processor capable of conference calling, call forwarding, automatic dialing, automatic redialing of busy numbers and last number redialing. Digital data networks have greatly increased access to on-line information retrieval systems and facilitated the transmission of large amounts of information between dispersed points. Facsimile transmission is widespread. Optical fibers can carry more simultaneous messages than conventional cable. Cable television has enormous potential for home information services, and videodisks and videocassette recorders could have a substantial impact on education as well as home entertainment. The publishing industry is being changed by the transmission of news and literature via telecommunications and broadcasting systems. Communications satellites provide the means for inexpensive, reliable and real-time information transfer on a global scale.

Recent Developments

It is the interactions among the functions of data processing, word processing and communications that reflect the real potential for advancement. Traditional distinctions between telephone utilities, newspaper and book publishing, banking and postal services are becoming blurred, and these blurred distinctions have generated important policy problems for both government and private enterprise.

In part, the issues have arisen because the electronic storage, manipulation, retrieval and transmission of information are available at costs competitive with paper. Several key examples illustrate the rapid development that has occurred in information technology in recent years.

- The cost of computer main memory has been declining 26 percent per year since 1965 and is expected to continue to do so through the 1980s.

- There is now a full range of commercially available storage technologies that permit access times as low as a billionth of a second for the small, high-speed storage used to process information.

- The new videodisks can store as many as 10 billion bits of information on a disk the size of an ordinary phonograph record.

- The performance of central processors has increased at a rate of 35 percent per year since they first were introduced (see Figure 3.1), while costs have declined by about 20 percent per year (see Figure 3.2).

- Circuit density is increasing dramatically, particularly in very large-scale integrated (VLSI) circuits. Integrated circuits now contain 100,000 active components; some estimates place the number at 10^9 by 1995.[1]

- Word processing devices are acquiring communications features that provide electronic mail functions as well as access to outside databases.

- Data processing, combined with word processing, provides a variety of information handling, storage and retrieval capabilities within one system.

- The use of microprocessors in a number of office devices – from dictation equipment to photocopiers – has reduced equipment size and enhanced operations significantly.

Donald J. Hillman

Figure 3.1. Relative Computer Processor Performance.

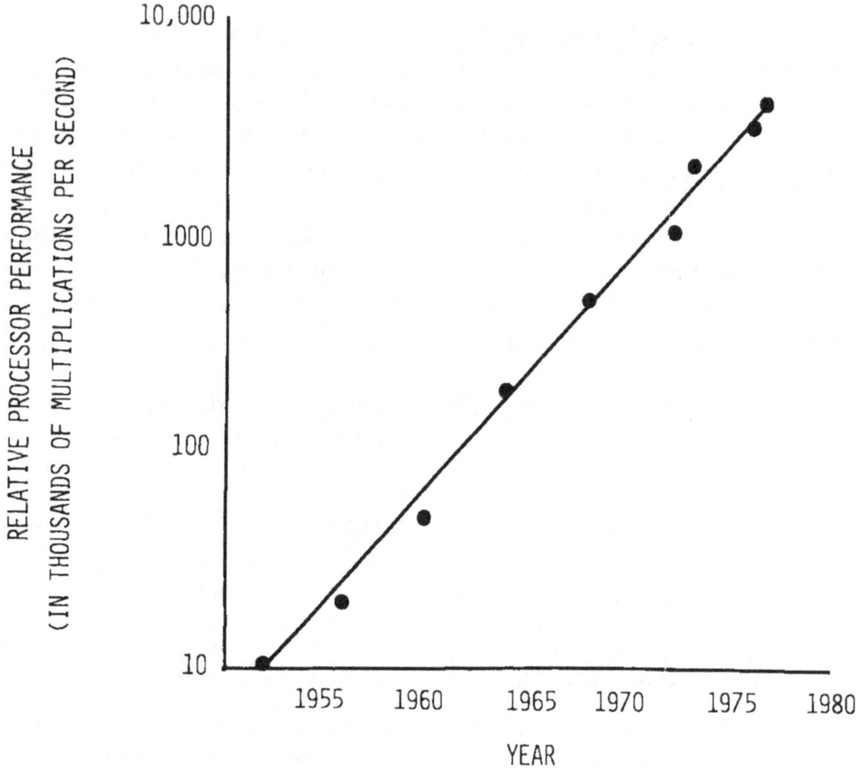

Source: Donald P. Kenney, <u>Microcomputers</u> (New York: Amacon Press, 1978).

- In the last 15 years, channel capacity of a single communications satellite has increased by a factor of 50, and the cost per circuit year has decreased by a factor of 45.[2]

- Packet-switching has become an alternative to time-division and circuit-switched networks.

- Some new services combine satellite, microwave and cable technologies for long-distance voice, data and video transmission.

Impact of Information Technology

The implications of these new tools are significant. Within the office, traditional operations are being altered as mid-level managers and technicians

Figure 3.2. Relative Computer Processor Price.

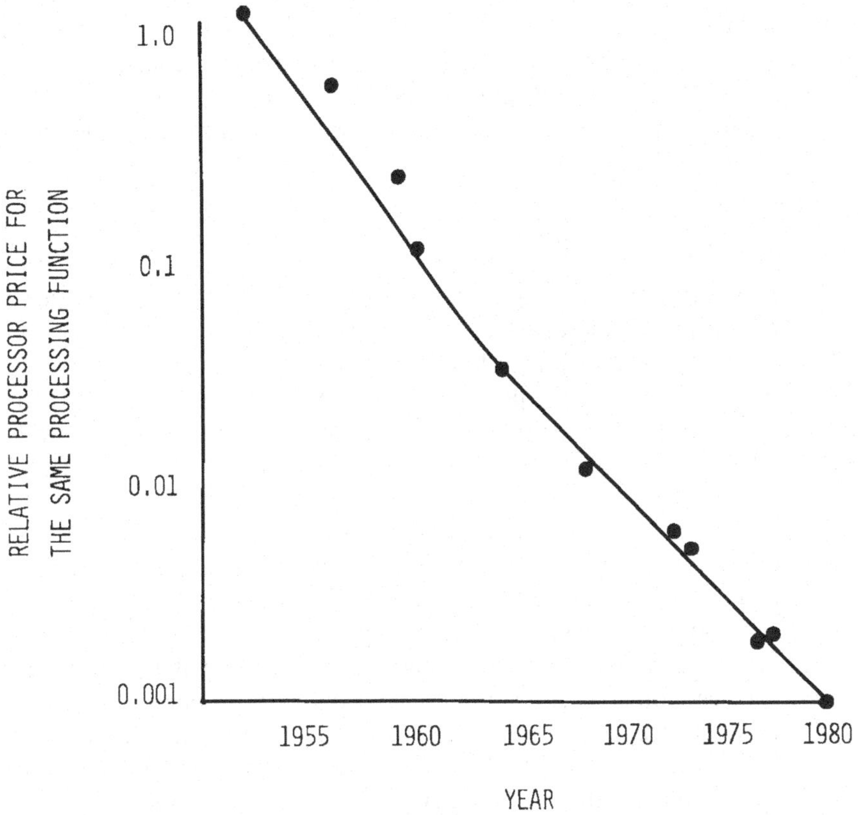

Source: Donald P. Kenney, <u>Microcomputers</u> (New York:
          Amacon Press, 1978).

employ minicomputers to monitor work routines. Writers and editors enhance their efficiency through the direct entry of data to word processors. Portable terminals enable employees to work from their homes. Line managers can receive timely information from a variety of distant sources, thereby improving the decision-making process. In some cases, the choice of technologies — for example, large central processing unit versus distributed minicomputers — will substantially affect the way an enterprise functions.

New information handling methods are affecting the commercial world in a number of ways, as in electronic banking and consumer purchasing via remote terminals. Multinational corporations operate more efficiently in the international marketplace through the use of communications technology linked to information processing.

Computer technology is entering the home as well. Microprocessors imbedded in a variety of devices control temperatures or turn on the oven. Innovative broadcasting and on-line services provide new forms of entertainment. Many home computers enable a user to balance a checkbook, play video games or take advantage of home education.

The revolution has brought us to the dawn of a so-called Information Age, whose implications for society are both disturbing and exciting.

The Rise of the Information Age

The role of information has increased with the growth of the new technology. The information industry is now the most rapidly increasing segment of the economy. The uses of information technology in all aspects of our lives have accelerated to the point where the 1980s can be called the Information Age. In recent years policymakers have turned more attention to issues related to information. Yet, despite the findings of numerous reports and studies, substantial issues remain unresolved.

In 1977, Porat[3] estimated that about 46 percent of the U.S. work force was employed in the information industry. This means that more people in the United States are employed in manipulating information than in manufacturing products, providing services or growing food. The Information Age is thus an "era in which the exchange of information will be as critical a function of economic organization as the production of goods."[4]

One consequence of this transformation is that information is now being valued as a critical resource in the same sense as is labor or capital. Information is different in that it conserves other resources through better decisions,[5] and it often is enhanced rather than depleted through use. Attempts have been made to treat information as a utility and to describe mechanisms for regulating it.[6] Another approach treats information as a mixture of purely public and private goods, with price reflecting an allocation mechanism rather than a cost-recovery device.[7] Common to all of these approaches is the goal of raising private and public productivity through improved information-handling methods.

The emergence of information as a tangible resource has stimulated a vigorous debate concerning its development. Do we need a national policy to manage our information resources? Is it possible to have a single national policy when information transcends so many activities and areas of government? While the recognition of the information economy is relatively recent, these questions have stirred debate in the United States for more than 20 years.

Numerous reports have addressed the issue of scientific and technical information collection and dissemination. Beginning with the so-called Baker

Report,[8] which recommended the 1958 formation of the Office of Science Information Service within the National Science Foundation, the role of information in research and development began to assume a more prominent position in the context of urgent national goals. In 1963 the reports of Wiesner[9] and Weinberg[10] promulgated the view that government was responsible for disseminating research results and for maintaining an adequate communications system to promote the commercial applications of those results. Another influential report, the report of the Committee on Scientific and Technical Communication (SATCOM),[11] recommended the participation of private organizations in the nation's information programs and suggested government support for scientific and technical societies.

In 1972, the "Greenberger Report"[12] expounded a global view of government responsibilities for information dissemination in research, education and private sector activities. Very much in the spirit of the times, the report championed the government's role in ensuring that the country's information resources were fully utilized for the public good. The report also recognized that centralization of the effort was not a necessity, and that the various public and private organizations engaged in producing and disseminating information were performing adequately, if not optimally.

Much progress has been made toward the goal expressed in these reports: the creation of a communication system for the free flow of scientific and technical information. Many observers, nevertheless, believe that these overall objectives have not yet been reached, partly because the constituency of users varies significantly, as scientific and technical information concerns merge with broader information policy questions. Throughout recent years, therefore, there has been a steady trend toward broadening the scope of policymaking for information, as seen in the formation of the U.S. National Commission on Libraries and Information Science (1975), the U.S. Domestic Council, Committee on the Right to Privacy (1976) and the U.S. Commission on Federal Paperwork (1977).

Despite these efforts, the tensions continue among various players in the federal government, and between the public and private sectors. The absence of clear guidelines and policies regarding information is keenly felt. It can be argued, for example, that the government should ensure the widespread dissemination of socially useful information at the lowest possible cost; on the other hand, some say that the private sector has the more efficient means for doing this. The issue often centers on the definition of unfair competition, especially with respect to the government's funding of information services.

A major reason for the conflict in the development of information policies is the absence of a suitable mechanism for resolving the issues. No distinct roles for the public and private sectors have emerged, and there is no agreement on planning and leadership. The decision-making process within the federal government concerning these questions is accordingly fragmented and disorganized. In response to this situation, there have been many proposals for new national structures to plan and coordinate information activities. Some have suggested that responsibility for the formulation of communications and information policy be centralized, either in a specific department or in the Executive Office of the President, although others contend that improved coordination efforts, together with a recognition that information issues play a large part in national policies, are sufficient to rationalize the decision-making process.

Both the private and the public sectors are deeply involved in generating and using information to manage a society that is increasingly dependent on problem-solving knowledge for a wide diversity of purposes and needs. Clearly, we need to focus more attention on ensuring cooperation both within the government and between the public and private sectors. This issue must be treated as a high priority question as the nation and the world grow increasingly dependent on the availability of information to provide answers to national and international problems.

Policy Questions of the Information Age

The Information Age has been evolving over a period of at least 20 years, as an economy based on industrial production adapts to one based on the transfer of information. The sheer speed of change is as significant as the changes themselves. The new issues created by the new technology affect all segments of society and include a broad range of problems, ranging from productivity to privacy to control of information production and manpower requirements.

Structure of the Telecommunications and
Information Industries

Telecommunications and data processing are merging as a result of the evolution of both technologies and the pressures caused by economic change. The blending of these two economic sectors into the critical component of the information industry is equally dependent on the ability of the technology to support this merger and the economic — and sometimes social — pressures for it.

The heart of the telecommunications network is now a computer — an electronic switch. At the same time, the usefulness of data processing facilities and services is a function of their accessibility through the telecommunications net. The growth markets for telecommunications and data processing exist in merged services, such as electronic message systems, electronic funds transfers and other transaction-oriented offerings.

Telecommunications and data processing borrow techniques from each other to increase the capacity and availability of existing services. Complementing these needs, however, are pressures for new services which could bring about increased efficiency and effectiveness. This second set of motivations is now the primary engine of change. While the reasons for viewing telecommunications and data processing separately appear to be fading, it is not clear that the two industries will be treated as one in policy and legal terms — at least for the foreseeable future. Unresolved policy choices concern regulation, competition and the assured delivery of services.

Regulatory strategy and requirements for the telecommunications industry are relaxing, but transmission services will remain regulated to a considerable degree, due to several factors. First, the industry is dominated by a corporation of unparalleled size — a circumstance that will not change quickly. Second, the technology employs an increasingly valuable and scarce resource, that is, the electromagnetic spectrum. Third, the industry must operate in conformity with certain powerful social and political policies, such as requirements for the universal availability of service.

A variety of other services will, however, be at least partly deregulated, as a result of Federal Communications Commission decisions, congressional legislation and state public utility commission actions. This process will affect the so-called enhanced telecommunications services, for example, and more participants will be able to enter the market. The challenge for policymakers will be to ensure that the transition is as fair and minimally disruptive as possible and to support continued technological advancement, rather than restrict it through burdensome regulations. Critical choices will emerge at the points where three distinct portions of the market meet: the unregulated data processing and information industries, the newly deregulated enhanced communications services, and the less (but still considerably) regulated transmission services. A key question will be that of how to draw the regulatory line between basic transmission services and enhanced offerings.

Information Overload

In the past, the information explosion was largely paper-oriented; today, the new technology is creating huge quantities of computer-readable material. The sheer amount of computer-readable data compounds the traditional problem of sifting useful information from a base of material that is uneven in quality. Reductions in the cost of storing, processing and retrieving information have only added to this dilemma. The result is that decision makers are often faced with the increasingly difficult task of selecting critical information from mountains of data.

A case can be made that technology can rationalize the information transfer process, thereby easing information overload. On-line searching, for example, provides improved methods for sorting and selecting needed information from large amounts of data. Systems using computer-based selective dissemination of information, which also highlight materials most closely reflecting a user's interests, eliminate the need to sift through endless computer printouts or printed documents. There is little doubt that as new techniques for data entry emerge—such as direct voice to computer—the amount of data gathered in computer-readable form will continue to increase at a dramatic rate. It is hoped that the technology can also provide the techniques to organize and display information more effectively, so that informed decision making will be enhanced rather than diminshed.

Privacy

The Federal Privacy Protection Commission carefully emphasized in its 1977 report that information privacy involves more than the traditional concepts of confidentiality imply. Personal privacy in this information society calls for fair practices in maintaining and using information, as well as restrictions on how organizations collect information. The driving force behind the concern for privacy is a desire to protect, not only the information about human beings, but also their autonomy and individuality. Contributing to this situation is the drastic reduction in costs for storing information in computers. As offices become highly automated, electronic mail and message systems increase and personal computers proliferate, the potential for abusing the confidentiality of information grows. Issues of personal privacy reflect another area where policymakers must address the adequacy of existing legal and institutional frameworks to cope with rapid technological advances.

Information Resource Management

There is a growing awareness that information transfer activities play a critical role in the effective internal management of both public and private sector organizations. As a result, information is increasingly viewed as an important resource to be carefully developed and utilized. In the United States no central authority controls the establishment or maintenance of information systems on a national basis. Recent efforts have been made, however, to coordinate government paperwork activities to reduce redundant collection activities and to increase the sharing of information.

Attention paid to this issue in recent years has grown as the number of databases increases (for example, there were 528 computer-readable, publicly available, bibliographic databases in 1979[13]), the variety of vendors offering information services expands and the methods for transferring information proliferate. This situation calls for effective coordination and management of all aspects of information gathering, processing, and dissemination activities within an organization to ensure optimal use of data resources.

This changing environment raises a number of policy questions, among them that of better defining the federal government's role in providing public information, in order to prevent conflict and competition between the government and the private sector. Other issues include improving access to vital information, increasing coordination of government information collection and dissemination activities and designing appropriate management tools and philosophies to improve information resource management.

Information Technology and Education

As the information economy permeates society, there is a growing need for professionals in engineering, programming and systems analysis beyond the number currently graduating from universities. Many observers believe that our educational institutions themselves should emphasize mathematics and science as preparation for careers in a variety of technological industries. Others contend that the federal government should provide increased funding to foster programs of this kind.

Information technology in education is also receiving renewed attention. Recent advances in miniaturization and telecommunications networks have opened up new ways of assisting students in a wide array of learning environments. They can optimize educational resources that may be geographically dispersed and offer new flexibility in individualized course work. Perhaps most importantly, computer and communications systems will help students to acquire skills for using these technologies throughout their lives and enhance their understanding of modern technologies in general. As the information society becomes more pervasive, the ability to employ automated systems for everything from commercial transactions to home entertainment will become increasingly important.

Availability of International Telecommunications and Information Resources

The growth of large-scale computer systems and telecommunications networks makes information available all over the world. As the globe continues to shrink, domestic and international activities and policies merge. The United States continues to be the leader in the field of information technology but is faced with an increasing challenge from its major trading partners.

Of particular concern are some nations' attempts to control the flow of information across their borders. The United States has traditionally supported the concept of free flow of information internationally as fundamental to world economic growth and human rights. Impediments to that flow could substantially damage the U.S. information industry through loss of exports. In the long run, other enterprises that rely heavily on information and communications products and services for efficient international operations will be most significantly harmed. Numerous noneconomic issues — such as national sovereignty, cultural erosion and personal privacy — are also linked to the international data flow problem. Several nations have responded to this problem by establishing national strategies and policies for information and communications development as well.

Radios are another of several communications resources that raise difficult questions. The radio frequency spectrum and the geostationary earth orbit are finite resources whose use is allocated by the International Telecommunications Union among its 154 members. As the spectrum becomes more congested, issues of equitable allocation of radio frequencies have become increasingly critical, as evidenced by the considerable attention focused on the 1979 World Administrative Radio Conference.

These international developments directly affect the ability of the United States to maintain its lead in numerous high-technology fields, as well as support the employment of information technology worldwide. It is unclear how the United States should respond and how the interests of domestic users of information technology can be best represented internationally. Other concerns focus on the complementarity of our domestic and international policies in this area and on the benefit at home of harnessing scientific and technical accomplishments abroad.

Information Technology Forecasts

To give a sharper edge to the issues raised above, I will make a number of forecasts of specific technological developments in the first half of this decade. These will be divided into products and services emerging from technological advancements, followed by a discussion of possible legislative and regulatory responses.

Information Technology Developments

To support the types of information transfer systems described earlier, different computer architectures will be required. Specialists are now developing the technologies needed for that architecture, including:

1. Hard software, which implements important software functions in specialized chips.

2. Very large-scale integrated circuits, which provide the foundation for the changes in technology and computer architectures.

3. Bus architectures, which link together hardware modules in computer/communications systems via standard interfaces, thus enhancing local area networks.

4. Extensible languages, which enable the language facility itself rather than traditional libraries to maintain extended functions.

Among the key components of these new architectures will be memory devices and database machines. With new memory devices—such as associative memories—data can be processed without first being transferred from slower memory to fast memory. Database machines approach the ideal of a plug-in database utility. These architectures are particularly important for very large database systems. As noted earlier, the number of databases is continually increasing. Similarly, the number of on-line searches has quadrupled to an estimated 4 million per year since 1975.[14] The market is expected to grow in response to the added value of databases as retrospective literature collections and as more students are exposed to on-line searching. This growth in database services will be possible as a result of two technological developments:

- new network architectures, which will provide superior access to stored information;

- mini-micro-based, on-line information retrieval systems.

These will enable end users to subscribe to customized portions of databases and to conduct all searches on an in-house mini-micro system.

These technological forecasts are by no means the only expected developments in information processing in the 1980s, but they are among the most important.

Emerging Information Services

The information services that will probably become widespread in the 1980s have, in many cases, already emerged. Several have been referred to earlier in indicating the scope of the communications revolution and the emerging policy issues.

Home Information Systems. Home information services will flourish in the next decade. Several experiments are currently testing consumer response to these systems, generally called teletext and videotex. Using different broadcasting or timesharing approaches, these systems supply information to the home television and in some cases provide for two-way communication. These systems offer a variety of services, including news, educational programs and consumer information, and will probably offer more in the future in response to rising customer demand.

Electronic Message Systems. Electronic mail systems are already operating within a growing number of private organizations. At issue is how such services might be made available to the general public. The U.S. Postal Service has developed what it calls Electronic Computer-Originated Mail (ECOM), but it is uncertain whether the Postal Service will be authorized to enter the electronic message business in competition with private sector enterprises. Whatever the outcome of this debate, the technology to accomplish electronic message transmission is readily available and increasingly part of home and office computer systems.

Electronic Publishing. Electronic book and newspaper publishing will flourish in the 1980s. Companies with profitable databases are even now seeking the technology to publish electronically. A number of traditional publishers are expanding to include capabilities for database publishing and delivery of home information services. One trade association estimated that 42 mergers took place in the first half of 1980.[15] This acquisition activity can be expected to continue during the 1980s, as corporate giants respond to the need for new technology.

Office of the Future. The office of the future will begin to take shape in the 1980s, as word processing and data processing functions are integrated. Future office managers will be able not only to create local communications networks but also to transmit large amounts of information over great distances. They will use the newly developed computer-generated graphics to display financial and operational data needed for management reporting and strategic analysis and thereby enhance their ability for informed decision making. A related activity in the workplace will be the widespread employment of computer-aided design providing impressive productivity gains.

Legislative and Regulatory Activities

At present, debate among Congress, the courts, the Executive Branch and the independent regulatory agencies is underway on several key policy questions. The growing complexity of these issues, combined with the long-range ramifications of decisions being made, requires continued analysis and concern on the part of policymakers. Among the core issues are the following:

1. *Regulation of the communications industry.* Should AT&T be permitted to offer information and data processing services? If so, under what conditions? What authority should the Federal Commu-

nications Commission have for regulating the communications industry? The data processing industry? Is legislation necessary or will anticipated court decisions settle the matter?

2. *Protection of intellectual property.* Both Congress and the courts are concerned about the effect of automation on traditional legal frameworks for protecting ownership of information. What should be done to adjust copyright provisions to a world of on-line databases and distributed computing systems? What protections should be awarded to software – copyright, patent or strictly trade secret?

3. *The role of government in providing information services.* Government and private sector vie as providers of information. What limits should be placed on the government to prevent unfair competition with private sector information providers? How can we improve coordination of public and private sector information needs and services? Should such activities as electronic mail and electronic funds transfer be left to the private sector, without government controls?

4. *Protection of personal privacy.* Growth of large computer systems and centralized databases will continue to spark concerns for individual privacy. Will Congress pursue additional privacy legislation in such areas as medical or insurance records? Will the greater use of computer systems in administering government programs require concomitant attention to computer systems security and confidentiality of records?

5. *Government organization of information activities.* This is one area likely to receive increased attention. How can the federal government improve its management of information resources? Is adequate government support being provided for research and development in information technology? In light of barriers to the flow of information being erected around the world, how should the government be organized to represent U.S. information and communications interests internationally?

The positive resolution of these questions during the next several years will require enlightened decision making. This necessitates a firm understanding of the state of communications and information technologies, of the significant advantages they bring to society and of the difficult policy issues they raise. A hard look by policymakers at the impact of new technologies on all sectors of the economy and the public is important preparation for the Information Age.

Notes

1. Howard L. Resnikoff and Edward C. Weiss, "Adapting Use of Information and Knowledge to Enhance Productivity," paper presented at the Conference on Productivity Research, American Productivity Center, Houston, Tex., 21–24 April 1980, p. 31.

2. Ibid., p. 23.

3. Marc Uri Porat, *The Information Economy: (1) Definition and Measurement; (2) Sources and Methods for Measuring the Primary Information Sector,* Department of Commerce, Office of Telecommunications Special Publication 77–12 (Washington, D.C., May 1977).

4. J. Becker, ed., *The Information Society: An International Journal,* announcement (New York: Crane, Russak and Co.).

5. Walter M. Carlson, guest editorial, *Information Manager,* vol. 2, no. 2 (1980).

6. Pranas Zunde, ed., *Information Utilities: Proceedings of the American Society for Information Science [37th] Annual Meeting* (White Plains, N.Y.: Industry Publications, Inc., 1974).

7. J. A. Dei Rossi, *A Framework for the Economic Evaluation of Pricing and Capacity Decisions for Automated Scientific Information Retrieval Systems* (Washington, D.C.: National Bureau of Standards, 1973).

8. U.S. President's Science Advisory Committee, *Improving the Availability of Scientific and Technical Information in the United States* (Washington, D.C., December 1958), p. 8.

9. U.S. Federal Council for Science and Technology, Committee on Scientific Information, *Status Report on Scientific and Technical Information in the Federal Government* (Washington, D.C., June 1963), p. 18.

10. U.S. President's Science Advisory Committee, *Science, Government, and Information: The Responsibilities of the Technical Community and the Government in the Transfer of Information* (Washington, D.C., January 1963), p. 52.

11. National Academy of Sciences – National Academy of Engineering, Committee on Scientific and Technical Communication, *Scientific and Technical Communication: A Pressing National Problem and Recommendations for Its Solution* (Washington, D.C., 1969), p. 336.

12. National Science Foundation, U.S. Federal Council for Science and Technology, *Making Technical Information More Useful: The Management of a Vital Resource* (Washington, D.C., June 1972), p. 59.

13. Martha Williams, "Database and Online Statistics for 1979," *Bulletin of American Society for Information Science,* vol. 7, no. 2 (December 1980), p. 27.

14. Ibid.

15. Anne Armstrong, "The Jonah Syndrome," *Bulletin of American Society for Information Science,* vol. 7, no. 2 (December 1980), p. 20.

4
Relations of Science, Government and Industry: The Case of Recombinant DNA

Charles Weiner

Introduction

The emergence and rapid growth of recombinant DNA and other new techniques for genetic manipulation posed major policy questions in the 1970s and will continue to do so as the research and its applications become increasingly visible in the 1980s. Several of the current issues were vividly highlighted by events during a five-week period in the fall of 1980 in Cambridge, Massachusetts, the center of the public confrontations on the safety of the research that made national headlines in 1976.

On 14 October 1980, a two-day biotechnology symposium opened at the Massachusetts Institute of Technology (MIT) to an overflow audience of 500, mostly from industry and investment companies. The first session featured five distinguished MIT molecular biologists. One of them noted in his introduction that many of the participants had been reading the *Wall Street Journal* that morning because, in a few hours, shares in a genetic engineering firm (Genentech) were to be offered for the first time on the open stock exchange. The speakers talked about the origins and principles of the basic science involved in recombinant DNA and cell fusion techniques, the frontiers of current research, possible applications and the difficulties that may be inherent to the science itself. Although enthusiastic about new applications, they emphasized that molecular biologists were only just beginning to understand the gene and its expression in higher cells. One urged the assembly of industrialists and potential investors to be careful not to "kill

Charles Weiner is professor of history of science and technology, Massachusetts Institute of Technology, Cambridge, Mass.

the goose that lays the golden eggs," warning that overly quick exploitation of the research could lower the morale of the scientific community.

At midmorning a speaker interrupted his prepared remarks to announce that the Nobel Prize in Chemistry had just been awarded to three scientists who had developed basic techniques for the new advances in DNA research and applications. During the coffee break, the participants were buzzing about another bit of news: The price of the newly offered Genentech stock had more than doubled within the hour. All of this sparked private dispute among several of the speakers about the appropriate relationship of fundamental biological research in the universities to the newly developing biotechnology industries. They were well aware of the issues; all five MIT biologists on the program were involved with commercial enterprises in the field.

Other dimensions of the new role of biology were brought out by events in Cambridge during the following four weeks. On 28 October, the Cambridge Biohazards Committee held a public hearing to gauge community reaction to the plan of Biogen, a major international genetic engineering firm, to establish a Cambridge facility for recombinant DNA research and manufacturing. Within a week, the Cambridge City Council reactivated the Cambridge Experimentation Review Board. This citizens' group had been created in 1976 to consider whether recombinant DNA research at Harvard and MIT posed a threat to the community and had gone out of existence in early 1977 after recommending what was to become the nation's first local ordinance regulating such research. In 1980, the council reconvened the board to consider Biogen's request to locate in the city. Two weeks later, in the city of Waltham (part of the Route 128 high-technology industrial area near Cambridge), the city council held a public hearing on similar issues related to the operation of a Waltham genetic engineering firm, Collaborative Genetics. In December 1980, Waltham approved an ordinance regulating recombinant DNA experimentation and use in the city. The ordinance requires such work to be done under the National Institutes of Health (NIH) guidelines and mandates some additional safety measures.

Meanwhile, a newly founded company, Genetics Institute, sought permission to establish its laboratory in Somerville, immediately adjacent to Cambridge. In January 1981, negative community response was expressed at a hearing attended by more than 100 people. One Somerville alderman challenged the credibility of a leading Harvard biologist as a safety expert because of his major role in the firm: "You're more than a scientist now. You're a businessman." [1] Harvard and MIT biologists are prominently involved in the three genetic engineering companies under public scrutiny by local governments in Cambridge, Somerville and Waltham.

Yet another indication of current reactions to issues raised by the applications of molecular genetics was the response by the faculty of Harvard University to the proposal that the university itself take a role in founding a company for commercial exploitation of recombinant DNA techniques, involving some faculty members and making use of university-owned patents. The intense discussions in Cambridge had echoes at other leading universities throughout the United States. Finally, in November 1980, the Harvard Corporation rejected the plan because of its potential for generating conflicts with academic values through interference with open communication, influence on criteria for promotion, intrusion on commitments to teaching and research and damage to the credibility and integrity of the university.[2] Harvard (and many other U.S. universities) continues, however, to explore ways of obtaining financial benefit from the applications of publicly funded research done in university laboratories.

These developments dramatically illustrate major unresolved national issues in contemporary relations among science, government and industry. Although the focus in this account is on recombinant DNA technology, similar issues arise in other areas of science and engineering. The problems involve:

1. The public's perception of new technology and determination of the role the public should play in defining the purposes and goals of new technology and the conditions under which it should be developed;

2. Determination of whether and how to regulate new technology;

3. The relationship of federally funded basic research to commercial exploitation of its applications;

4. The effects of increased university interaction with industry on the direction and quality of basic science, on the community of scientists and on the university environment.

In addition, ethical issues stemming from the applications of genetics will be increasingly important in the 1980s, at the level of practical reality rather than mere abstraction.

This paper reviews the background and current status of these issues in relation to recombinant DNA techniques. Because much has already been written about them, I will emphasize only those issues that have not yet received adequate attention (see the bibliography that follows this paper). The field and the policy issues are unfolding before our eyes, providing an opportunity for close observation of the social and intellectual processes of the growth of knowledge and its uses, and emphasizing the need for critical examination of the related policy processes.

Growth of Recombinant DNA Research and
Concern about Risks in the 1970s

The development of recombinant DNA techniques in the early 1970s was a major event in the history of science and a result of decades of fruitful research in molecular biology in several nations. These techniques involved the use of newly discovered restriction enzymes to isolate and remove specific gene sequences from DNA molecules of various organisms and to recombine them with the DNA of other organisms. These techniques also involved the application of methods to reproduce large amounts of exact copies (clones) of the hybrid or recombinant DNA molecules. The ability to manipulate nucleotide sequences directly made it possible to transmit genetic information among different species. It provided a powerful new tool for study of the structure and function of genes and made it possible to study details of DNA and its transcription in cells of higher organisms. Biologists immediately recognized the major significance of these developments. They were now able to solve problems at the forefront of knowledge, which also had possible important applications.

Of comparable importance was the fact that the scientists involved in the work made an unprecedented effort to inform the scientific community—and, indirectly, the public—not only of the exciting new advances and potential benefits of their basic research but also of their concern about potential hazards in their own laboratory work. The extraordinary extent of these concerns was made visible in 1974 when these researchers called on fellow scientists to refrain voluntarily from doing certain experiments until their hazards were assessed and safeguards devised. From the start, throughout the February 1975 Asilomar conference, and during subsequent efforts to develop guidelines, the issue was defined by the scientists as a limited problem that the scientific community could solve by technical means. Possible abuse or misuse of the research was mentioned occasionally, but it was excluded from major consideration, as were also its social implications. The discussions focused on whether there was any danger in the research and, if so, how the danger could be avoided while the research continued.[3]

Scientists, sensing the excitement and popularity of the new field, were eager to do recombinant DNA research. Developers of the guidelines, convinced the field should be allowed to grow, were reluctant to impose artificial restraints. The National Institutes of Health were thus in the ambiguous position of encouraging the growth of research using recombinant DNA techniques while taking responsibility for establishing and enforcing safety regulations restricting such research. Concurrently, scientists were tooling up to use the new techniques and were waiting for the green light to proceed as rapidly as possible. Many of those charged with developing guidelines felt that, even though they lacked information, they had to move as quickly as possible.

The scientists involved believed that they needed to demonstrate that scientists, on their own, could act responsibly to protect the public. They felt that if they didn't do it on their own, someone else would do it for them. When pressures concerning regulation exist from both scientists and the public and a large degree of uncertainty prevails, there is bound to be disagreement on the scientifc basis of risks and the weight attached to them. The process of establishing rules therefore involved formulating a series of compromises. It was necessary to provide a framework for safe conduct of the research acceptable to the scientists affected by the regulations and, at the same time, to assure the public that it would be protected against possible hazards. The NIH Guidelines for Research Involving Recombinant DNA Molecules took effect in July 1976 and subsequently governed federally funded university research in the United States. They became the model for guidelines in several other countries as well.

By 1977, scientists in the field were issuing public statements explaining that they now felt that the hazards had been exaggerated and that, in fact, the experience of biologists since 1973 (when they first sounded the alarm) had convinced them that much of the concern was groundless. These statements came after a year of public controversy regarding the risks involved and after Congress had begun to consider legislation to regulate recombinant DNA research. The experience of the biologists since 1973 had been political as well as scientific.

Public interest had been relatively limited until 1976. Then, as research began, concern surfaced in several academic communities. Some public controversy was sparked by scientists who were critical of the guidelines on scientific grounds. But many nonscientists quickly realized that the information essential for evaluating potential mishaps (and thus, the adequacy of the guidelines) was not available. There were many unknowns. Were the guidelines adequate? Were they to be believed as a matter of faith? Which scientists should be trusted under such circumstances? Were scientists acting out of self-interest when they assured the public that research was safe? What public health benefits might be delayed or lost by slowing down the research until more was known about the risks?

By the end of 1977, 16 bills were introduced in Congress and the subject was widely probed in 25 hearings. At first, many scientists were prepared to accept the inevitability of federal legislation, which they hoped would extend NIH guidelines governing academic research to industry and prevent the proliferation of local regulations more severe than the guidelines. They soon mobilized vociferous opposition to legislation that they feared would be too rigid for a new field in which the perceptions of risk were changing rapidly. They believed that such legislation would restrict their research and threaten their relative autonomy. In their lobbying efforts, the scientists argued that new scientific evidence and analysis of existing data demonstrated that the probability of risk from recombinant DNA experiments was much lower than they had originally thought. Sympathetic media coverage and increasing references to impending medical benefits of the research contributed to a changed congressional mood. None of the pending bills came to a vote.

Rapid growth of the research and its applications occurred despite the public disputes, restricting guidelines, special containment facilities and related bureaucratic impediments that reached down to the laboratory level. While the regulatory issues were being debated in committees and in public hearings during the late 1970s, the laboratory use of recombinant DNA techniques was booming, as its potential was being explored in one subfield of biology after another. It became a central tool for research that previously had been considered impractical or impossible. Combined with other new developments, such as rapid methods of DNA sequencing, it has created great intellectual excitement and activity affecting all of biology, including cell biology and immunology. Already the research has led to a dramatic change in the understanding of the structure of genes through the discovery of intervening sequences and greater understanding of transposable elements.[4] Recombinant DNA methodology rapidly has become a required technique in molecular biology laboratories, and more and more scientists are using this powerful approach because of its simplicity, its effectiveness and its fruitfulness in opening new areas of research. The increase in the number of federal grants, the emergence of new journals and newsletters, special conferences and training workshops devoted to recombinant DNA research all indicate the enormous growth of research in the field.[5]

Although potential applications were clear from the beginning, they became more apparent as the research progressed and developed much more rapidly than had been anticipated. Recombinant DNA and gene sequencing techniques have made it possible not only to isolate and analyze genes but also to engineer genes to make specific proteins for synthesis of such substances as insulin, somatostatin, interferon and other polypetides with important biomedical application. In addition to pharmaceutical and medical applications, the use of the technique in producing industrially useful enzymes has also sparked great interest, and activities are underway in the agricultural, chemical and energy areas. By 1980 an estimated 100 U.S.

companies were evaluating or conducting recombinant DNA or other biotechnology research.[6]

On balance, it may well be that concern and controversy over risks and the need for control of research have accelerated growth of the field rather than retarded it. This hypothesis needs further study, but it is supported by several consequences of the special treatment given to recombinant DNA research. From the beginning, a need was perceived to assess the nature and potential of the research in order to determine whether it posed risks and, if so, to devise methods to reduce and contain them. This led to acceleration and supplementation of normal channels of scientific communication. Highly publicized meetings such as Asilomar in 1975 acquainted many scientists with the background of the research and with the newest results considerably earlier than they otherwise would have learned of them. The development of guidelines, the establishment in 1976 of the NIH Office of Recombinant DNA Activities, which distributed information on safe and efficient host-vector systems for research, and the publication of a newsletter were among the institutional efforts that stimulated the informal communication network in the budding field and enlarged its scope.

In addition, the discussions of the risks and benefits of the research in public forums, hearings and the media stimulated interest among investors, industrialists and scientists in the possible applications. A number of biologists never before involved in applied research began to consider commercial uses of their work. In some instances, the early results of such efforts to produce substances with important human medical applications were reported directly to the press and to congressional committees engaged in legislative hearings even before they were published in scientific journals, in order to bolster the argument that the benefits side of the research outweighed the risks.

All of these activities contributed to the growth of a technique that had great scientific merit from its inception because it had strong intellectual appeal, provided fruitful opportunities for research and publication and was relatively easy to learn and do. Some experiments and some applications, however, were temporarily delayed because of the time involved in the process of establishing, revising and interpreting safety guidelines. In addition, some scientists in the field had less time for research because of their participation in committee meetings and public hearings and because of the increased paperwork related to their laboratories. The overall cost of the research was increased by the guidelines and risk-assessment activities and by the expenses for new containment facilities, many of which are now no longer required because of changes in the guidelines. Funds for risk assessment and new facilities usually came from budget categories designated for evaluation or building and did not drain funds available for research. On the whole, these delays, distractions and costs appear to have been more than offset by the acceleration of the field caused by the extraordinary diffusion of information and the intrinsic appeal of the research.

Current Status of DNA Technology and Its Regulation

What is the status of regulation, risk assessment and public perception and involvement in recombinant DNA technology at the beginning of the 1980s?

Regulation

The NIH guidelines are the only regulations specifically applying to recombinant DNA research. They have been adopted by other federal agencies and now are mandatory for all federally funded research. Noncompliance can result in withdrawal of funding from the institution. Primary responsibility for determining that experiments are carried out in accordance with the guidelines is assigned to the Institutional Biosafety Committee (IBC) at the institution where the research is done.

The guidelines have undergone three major revisions by the NIH Recombinant DNA Advisory Committee (RAC) since 1976 and are continually subject to revision.[7] The trend has been to relax physical and biological containment standards and accountability protocols. At present, about 90 percent of recombinant DNA work being pursued in the United States is either no longer covered by the guidelines or subject to only minimal controls equivalent to "standard laboratory practice." In most cases researchers do not need to use the safety systems that had been introduced for biological and physical containment under the original guidelines. NIH further reduced its oversight role in November 1980 by eliminating the requirement for researchers to register and receive NIH approval before initiating experiments for which the guidelines already specify the containment level. This responsibility is now in the hands of the IBC at the institution where the work is to be done.

NIH has not yet studied the effectiveness of the IBCs but plans to do so. When the heads of the IBCs from almost 200 institutions met in Washington late in November 1980, many balked at the added responsibility of such an evaluation. Many participants believed that the IBCs could not be justified exclusively on the basis of the potential risks of recombinant DNA research. They believed that such research posed no greater hazards than identifiable biohazards in other fields.[8]

With the proliferation of industrial activity in the field and in the absence of federal legislation, NIH has developed procedures for voluntary compliance with the guidelines by industries using recombinant DNA techniques, and most, if not all, of the companies have announced that they will comply. The RAC is currently debating its role in regulating industry through voluntary compliance. Many members and outside observers have challenged the ability of the committee to make judgments and take responsibility regarding industrial practices where they have inadequate expertise to deal with large-scale fermentation processes and no authority to monitor or enforce compliance with the guidelines. The trend has been for the committee to recommend reduction of its responsibilities for the safety of research in the private sector, even in the absence of evidence that the regulatory agencies are playing an active role in the field.

Several federal agencies are currently considering their roles and are participating in the industrial practices subcommittee of the Federal Interagency Advisory Committee on Recombinant DNA Research, and a few are beginning limited efforts at regulation. The Food and Drug Administration is in the final stage of defining the policy process to regulate drugs produced by recombinant DNA techniques. The National Institute for Occupational Safety and Health has initiated a series of on-site surveys of companies starting up large-scale recombinant DNA work and is studying appropriate recommendations to industry for medical surveillance of employees. The Environmental Protection Agency has established a research program to provide a database on the environmental impacts of large-scale genetic engineering, including studies of the establishment and persistence of novel genomes in a variety of environments, modeling of the probability of escape of organisms from containment and exchange of genetic information.

In addition, the Office of Technology Assessment has completed a study called *Impacts of Applied Genetics: Micro-organisms, Plants, and Animals,*[9] which reviews several aspects of the subject, including current regulatory activities, and formulates options for congressional action. A Senate oversight hearing on industrial applications of recombinant DNA techniques was held in May 1980, and, although a bill (S. 2234) to register all recombinant DNA research with the Department of Health and Human Services was introduced earlier in the year, it remained in committee and no further action was taken by the end of the Ninety-sixth congress.[10] There is little evidence of widespread congressional support for special regulation in this field.

Risk Assessment

Potential safety risks of recombinant DNA research have been the focus of concern since the early 1970s. In 1977 and 1978 a consensus emerged among researchers in the field that the potential risks were less serious than had been originally feared. Upon more reflective analysis of existing data, these biologists became convinced that most of their original concerns were unfounded. At conferences and workshops in 1977 and 1978, efforts were made to assess risks on the basis of available knowledge. The results were reassuring to the scientists and encouraged them to relax the guidelines.

At that time, however, the first experiments specifically designed to assess risks in this field were just getting under way. Defining experiments to assess risks in a rapidly changing new field had inherent difficulties, and efforts of this kind lacked precedent and experience.[11] Risk assessments had been further delayed by lack of interest among scientists and by bureaucratic obstacles. Some of the biologists involved in the development of NIH guidelines maintain that risk assessment was undertaken reluctantly, in response to political pressure, rather than wholeheartedly, in response to technological reasoning.

Others argue that, although some questions have been answered, there is still too much uncertainty and not enough systematic knowledge of risk assessment. Several thoughtful risk-assessment experiments have focused on specific areas where information was needed. Although these experiments have laid to rest a number of the concerns that had been raised, some researchers in the field and some scientists on the RAC feel that the interpretation of data from these experiments has been overly optimistic and that problems and ambiguities noted by the investigators have been overlooked or unduly minimized. In some cases, they argue, generalizations have been prematurely made at a stage when scientific knowledge and prudence call for further case-by-case analysis. They maintain that too few risk experiments have been done and warn against premature abandonment of risk assessment.[12]

The development and annual update by NIH of a comprehensive risk-assessment program was mandated by the Secretary of the Department of Health, Education, and Welfare when the revised NIH guidelines were issued in December 1978. It was not until late 1979 that NIH, the agency responsible for both the promotion and regulation of the research, published the final version of the first plan. The NIH proposal for the first annual update of the plan was released for public comment in September 1980, and the final version of the revised risk-assessment program was announced in June 1981.[13]

The hazards of recombinant DNA research remain hypothetical after five years of intense research conducted under safety guidelines at laboratories throughout the world. The new knowledge gained from the research, the promise of its applications and the absence thus far of demonstrated risk have all contributed to a lack of enthusiasm among researchers for vigorous risk assessment. Now that some political battles have been won and the public mood seems favorable, much of the will to devise and conduct risk studies has disappeared. Researchers who have lingering doubts about the

safety of specific experiments or the adequacy of the containment prescribed by guidelines are reluctant to discuss them publicly, because they do
not want to be labeled as dissidents or to unleash new negative public reactions. Some researchers are critical of those scientists who, in their
eagerness to reassure the public, rashly proclaim that the research is perfectly safe. These enthusiasts, some argue, may provoke a backlash if there
is a real (or perceived) mishap. The very success of recombinant DNA
research in yielding knowledge about genetic structure has had a sobering
effect on many scientists, who realize that the field is full of surprises.

At the same time there is apprehension that large-scale industrial operations might pose special problems for risk assessment. In some communities, continuing doubts about the safety of recombinant DNA research have
been coupled with suspicions about the responsibility of industry generally,
especially in the wake of increasing public awareness of occupational safety
and industrial toxic waste problems. The motives of scientists who offer
reassurance have also been questioned because of their assumed financial
stake in the outcome. Federal and state regulatory agencies have done little
to address these local concerns because they are reluctant to step in when no
risk has been demonstrated and they lack the appropriate expertise to assess
risk themselves.[14]

Public Involvement

By now it has become clear that the question is not whether the public
should participate in scientific and technological affairs that have important
social consequences, but how they can participate effectively and intelli-
gently. Despite high public interest in the new developments in molecular
genetics, opportunities for public participation at the decision-making level
are still limited and participation is often ineffective. Opportunities for
public input at the 1976 and 1977 NIH hearings on the guidelines created
some channels for public comment, and NIH has published an extensive
record of the guidelines process.[15] More recently, the public has shown little
interest in the guidelines on the national level, and the media have only
sparsely covered them, except when they have been reportedly violated by
researchers.

Several positions on the NIH Recombinant DNA Advisory Committee
have been designated for "public members" (about one-third of the present
members, including the chairman, are not scientists). Their participation
has introduced some policy issues that otherwise might not have been
raised.[16] Yet most of the issues placed before the committee are technical
and generally beyond the expertise of members not trained in the relevant
scientific fields. Many of the scientists on the committee have made special
efforts to explain technical matters to nonscientist members. Several of the

nonscientists have developed considerable ability to discuss many of the technical issues. Dissenting views on matters of procedure and values—most recently regarding industrial applications—have been regularly advanced by several public members, but they comprise a relatively isolated minority on the committee and generally have been heavily outvoted. The public members have not tended to vote as a bloc. Although the committee's meetings are open to the public, most of the observers who have attended during the past several years have been representatives of industrial firms with interests in the field.[17]

A few groups, such as the Cambridge Experimentation Review Board, have been founded, but they have been short-lived and have not been evaluated fully.[18] They were initiated on short notice in response to a crisis and never developed a continuing involvement of the public. The reactivated Cambridge board, whose membership was virtually the same as that of the original board of 1976, was able to build on its past experience and function more effectively. In assessing the adequacy of regulation of industrial recombinant DNA activities, the board consulted its own experts from a variety of relevant fields, including authorities on fermentation processes and sewage disposal.

Applied Molecular Genetics in the 1980s: Policy Prospects and Problems

Public Expectations

As the public controversy over the risks subsided in the late 1970s, attention focused on the benefits of the research. In 1980, a steady stream of enthusiastic accounts in the scientific, business and popular press, and a number of workshops and conferences hailed a revolution in molecular genetics and the birth of the Age of Biotechnology.[19] This exuberant response was based not only on the power of recombinant DNA technology but also on its appeal as an embodiment of current values. Recombinant DNA is presented as a panacea that will increase productivity and profit, help solve the energy problem, increase world food production and improve the public health. It is also regarded as consistent with the need for protection of the environment and conservation of resources. The new genetic technology is promoted as a "tech fix" for lagging rates of economic growth and productivity. For those seeking solutions to economic problems through technological innovation and transfer, recombinant DNA techniques represent ideal examples of successful experiments in this direction. Some academic institutions advance similar arguments in the hope that applications of genetic research developed in university laboratories can provide them with needed income.

The enthusiasm also reflects the desire to demonstrate to the public that the research has beneficial applications and to discourage unwarranted fears that might needlessly delay bringing needed products on the market. In addition, recombinant DNA technology is offered as a dramatic example of the ultimate payoff of basic research. Government policy has been based on the assumption that even basic science research should pay off if it is to receive public funds. Historically, this assumption has led to pressure on scientists for visible and immediate results. The clear potential of biomedical research to alleviate human suffering places it under special pressure. The history of the relationships among Congress, the National Science Foundation, NIH and other executive agencies in the past two decades demonstrates that the political environment significantly influences the establishment of research priorities, tending to favor areas such as cancer research that may yield results of current national interest.[20]

The new industry of genetic technology hopes to develop, manufacture and profitably market needed products through the exploitation of state-of-the-art genetic engineering techniques. Although the prospects appear promising, the technology is largely untried and the scientific understanding basic to it is largely incomplete. The effectiveness, safety and economic advantage of the products have yet to be demonstrated. Early publicity about human therapeutic substances, such as human insulin, growth hormones and interferon—all produced using recombinant DNA techniques—has not always made clear whether the gene product was biologically active and performed the same functions that it would perform naturally.[21] The genetic technology industry must purify its products to separate them from unwanted or unsafe substances that might be produced by the bacterium containing the recombinant DNA molecule, and all of this has to be economically feasible. Because tests are necessary to establish the safety and efficacy of the substances, there may be considerable time between laboratory research and commercial availability. Competition in the biotechnology industry is hastening the pace, however; a few companies have already started human testing of bacterial insulin, growth hormone and interferon.

In raising hopes of solutions to major health problems, the genetic technology industry may be overselling the public on the new technologies. For example, magazines and newspapers have already described interferon as a cancer cure. In this situation natural human optimism is exacerbated by the highly competitive nature of the industry, the prospect of large profits, the predictable enthusiasm of pioneers opening up new fields, the proliferation of new companies with a need to attract investors in order to get off the ground and the idealization of this new technology as a solution to economic, social and health problems. Unfulfilled expectations might well lead the public to doubt the credibility and motives of scientists and to become disappointed and impatient with the pace and direction of research. Discontent might also develop among young scientists recruited into the genetic technology industry if the prospects for career development and continuity do not materialize in what originally appeared to be a glamorous, intellectually stimulating and lucrative field.

Just as it would be irresponsible to overstate the claims for the new genetic technology industry, so would it be unsound to encourage and facilitate its growth without careful consideration of important unresolved policy issues concerning the relations of science, government and industry.

Patenting of Living Organisms

The June 1980 Supreme Court decision (five votes to four) permitting the patenting of "a live human-made microorganism" opened the door to action on more than 100 patent applications based on recombinant DNA techniques. The majority opinion of the Court took the position that the distinction between living and nonliving things was not relevant to the granting of a patent, and that the criteria for issuing a patent must rest on whether the genetically manipulated bacterial strain was "a product of human ingenuity" or "a product of nature." This opinion rested on interpretation of the intent of Congress as expressed in the Patent Act of 1790 (embodied in 35 U.S.C. Section 101), the 1930 Plant Patent Act and the 1970 Plant Variety Protection Act. The Court stated that Congress should debate this question if it disagreed with the ruling. The dissenting opinion held that Congress had not foreseen the new areas made possible by genetic engineering and argued that Congress must act before the Court could extend patent rights into such areas.[22]

The decision received wide press coverage and stimulated discussions about the ethical implications of private ownership of life forms. The impact of the decision on the genetic engineering industry and on the free flow of scientific information was also considered. The Supreme Court said its decision rested on a narrow interpretation of patent law, and some observers subsequently argued that the ruling did not involve important larger issues of public policy. Others maintained that the Court was indicating that the legal basis for its decision was inadequate and was inviting public bodies to prepare and discuss legislation.[23] Congress can, of course, enact legislation to prohibit patenting of living organisms, whether they are modified or not, or it can specifically provide for patents of living organisms to whatever extent it sees fit.

Congressional hearings were held on the subject during 1981, and the

President's Commission for the Study of Ethical Problems in Medicine and Biomedical and Behavioral Research is examining the issue as part of a more general study on genetic engineering. A "Public Forum on Patentability of Microorganisms" was held in July 1980 by the American Society for Microbiology and the House Committee on Science and Technology, but there has as yet been no other visible activity in Washington.

Public interest in this issue is high, and additional opportunity should be provided for public discussion of appropriate ways to deal with the development, ownership and use of living organisms. Questions that have not been adequately considered are:

- What effect will patenting have on the overall development of biomedical research? Does genetic research have special problems not shared by other fields? What can be learned from the history of the effect of patents in other areas of research?

- Who should profit from commercial applications of publicly funded research? Private industry? Scientists whose research yielded the applications? The academic institutions that sponsored the research? The citizens whose tax dollars supported the research?

- What ethical considerations should be taken into account when deciding patent policy? How can effective public input be obtained, and what role will it have in the formulation of policy?

Many scientists are deeply concerned about the threat that commercial interests may pose to the traditional free exchange of data and to open publication in scientific journals. Peer review, verification of results and, ultimately, the growth of knowledge are not possible when research procedures are kept secret for commercial reasons. Even though the patent laws require considerable disclosures, many scientists are concerned that the rapid industrialization of newly spawned basic research will skew the intellectual development of the field and will degrade cooperation within the scientific community. Even before the Supreme Court decision, academic biology departments were disturbed by the possibility of commercial gain, which sparked disputes among colleagues, aroused suspicions of piracy and premature publication and interfered with the exchange of data, bacterial strains and cell lines.[24]

Stanford, Harvard, Yale and the University of Michigan are only some of the universities investigating ways to retain an interest in potentially profitable patents. These institutions are, by and large, responding to increasing involvement of university biologists with private companies, either as consultants or founders and part owners. Individual scientists are voicing their concern about the effect of commercial interests on their field or on their institutions, in private discussions or through group letters circulated among their colleagues. Yet the scientific community and the public have had little opportunity to discuss these issues systematically. Assessment of the effect of existing arrangements on the university, on the health of science, on industry and on the public has, to date, been inadequate.

The Health of Science

The recombinant DNA case directly involves three major factors that contribute to the health of science: the strength of the universities, financial support for basic research and the social system of the related scientific community.

Major research universities, where most of the work in basic science has traditionally been done, increasingly complain of impending financial shortfalls because of the steadily rising costs for plant, equipment and personnel, especially for costs associated with scientific research. For several years, university administrators have warned that federal support is not keeping pace with the increase in operating expenses required to maintain high standards of research. In addition, federal support for basic research is decreasing (in constant dollars) in some fields. Government agencies that traditionally have supported all or most basic research in certain fields are increasingly under pressure to emphasize practical results, and many are trying to hasten the transfer of scientific knowledge to practical technology.

Influenced by the commercial applications of DNA techniques, a number of university researchers, not previously involved with industry, have become industrial consultants, joined industrial laboratories or taken leading roles in founding new companies. The research on which the applications are based was developed primarily in university laboratories supported by public funds. In past attempts to reap some of the financial benefits from new developments, major universities developed a variety of arrangements to benefit from the ownership and licensing of patents. (For example, the Wisconsin Alumni Research Foundation was established in 1925.) Several universities are currently discussing other arrangements to retain a portion of the profits generated from university research.

The proposal recently advanced by the administration of Harvard University provoked strong objections from its own faculty and was withdrawn, but the issue at Harvard is by no means settled. The proposed Harvard experiment suggested major revision of the university patent policy and alteration of the formal relationships of the university to its investments and of the faculty to industry. It would have involved the university and some of its biology faculty in founding a genetic engineering company in which both the faculty members and the university would be shareholders, along with private venture capital investors. The university, as owner of any patents resulting from work in a professor's laboratory, would license use of those patents to the company. At one point in the discussions, the Harvard administration proposed that space in a new biochemistry and molecular biology building be used as temporary quarters for the company.[25]

The reaction of the faculty, first in the biology department and then throughout the university, was overwhelmingly negative. The proposal was discussed at faculty and department meetings and in group letters circulated within the university. It stimulated comments in the national press.[26] Despite the university's claim that there would be safeguards to prevent abuse of the system, opponents argued that major university investment in the commercial work of faculty members would compromise academic freedom and lead to unavoidable conflicts of interest, to the detriment of the research and educational responsibilities of the faculty and the university.

Ultimately, the Harvard Corporation, stating the "academic risks outweighed the financial gain," voted to withdraw the proposal.[27] Explaining the decision, President Bok cited several of the objections that had been raised including (1) that academic discussion could be impaired because of commercial competition; (2) that professors and graduate students might shirk academic duties and interests to pursue commercial ones; (3) that the administration's authority to protect its academic interests might diminish; and (4) that Harvard's reputation for academic integrity might be damaged by even the appearance of conflict between its academic and financial interests. The university administration emphasized, however, that it badly needed additional sources of funding to strengthen the university's teaching and research and that it would continue to explore similar proposals.[28]

For several years, Congress has discussed the appropriate relationship of the university to commercial exploitation of federally funded research done on university premises. Several recent congressional actions focus on transfer of technology and university licensing of patents and will probably stimulate great interest in the near future. Congress has not, however, adequately considered the need to provide stable and increasing support for basic research. Nor has Congress recognized the importance of protecting the university from damaging pressures that would impair the quality of research and inhibit open communication among scientists.

These problems were stressed in November 1980 when a group of recent Nobel Prize winners visited the House Subcommittee on Science, Research, and Technology to appeal for more funds for basic research and for greater congressional sensitivity to the special problems of the scientific community. On this occasion, Hamilton Smith, the microbiologist who shared the Nobel Prize in 1978 for work that laid the foundations for recombinant DNA research, expressed his concern that the rush toward commercial applications in biology would harm the academic environment that has nurtured basic research. Smith noted that "free exchange of scientific information . . . may suffer, and long-term progress may be traded for short-term financial gain. . . . We still do not know the structure of human chromosomes, how the genes are arranged, how tissues and organs are formed, or even how any single human gene is regulated and expressed." Smith called for increased federal support of such academic research to "prevent the gutting of the university faculty" by new companies in the field.[29]

An earlier warning had been sent to Congress in 1978 by another Nobel laureate in biology, Joshua Lederberg, who predicted, "The possibility of profit — especially when other funding is so tight — will be a distorting influence on open communication and on the pursuit of scholarship."[30] Lederberg wrote that he did not think that his views were widely shared within the universities. In 1981, however, these problems have developed into a major concern.

University scientists in some fields of physics, chemistry and biology have long been involved with commercial applications of their research, especially since the end of World War II. However, there has been little systematic evaluation or historical analysis of the effects of these experiences on the university, on the research environment, on the direction and quality of basic science or on the scientists themselves. Understanding of these effects — and how they have differed for specific scientific disciplines (and groups within them), institutions and historical periods — would be especially helpful in assessing and responding to the rapid changes now underway in molecular genetics.

The events of the past decade — the development of powerful new research techniques, the demands for increased public scrutiny of the procedures and goals of basic research in molecular biology and the new relevance of such research for industrial and biomedical applications — have had profound effects on the community of researchers involved. The excitement of these scientists over the possibility of opening up new frontiers was coupled with concern that safety problems and public distrust might hamper the research. The regulatory procedures, public confrontation and political battles were new and unexpected, and the rapid growth of opportunities for commercial applications of their work raised new dilemmas. Solutions to the current problems must take into account the effects on the health of the scientific community.

Social and Ethical Consequences

The enormous potential of genetic technology in a variety of fields has been much heralded. Even if only some of the hopes of its promoters materialize, the new technology will surely transform our lives in the next decades. Despite the highly visible adverse effects of technology in recent times, there has been little public discussion of the potential economic, social and environmental consequences of new technologies, nor has there been debate on desirable priorities for application. (Belatedly, energy technologies are now under debate.) Biotechnology presents an opportunity for just such constructive discussion and planning. Which applications are socially valued? Which may be undesirable? What would we like the technology to do? Does it automatically serve "human needs" and "public purposes"? Who should decide about its uses, and who will benefit from it? Can we in good conscience introduce and encourage the growth of a powerful new technology without asking why we are doing it and for whom?

More than a decade ago the influential technology assessment report of the National Academy of Sciences (NAS) emphasized the need to pose such questions at an early stage in the development of new technology. The NAS report stressed that, in decision making on technology, a wide range of human values and concerns should be considered, policy options should be preserved and efforts to reduce uncertainties should precede or accompany decisions. The report called for favoring technological projects or developments that leave maximum room for maneuver and noted that "the reversibility of an action should thus be counted as a major benefit; its irreversibility, a major cost." It also called for limits "on the extent to which *any* major technology is allowed to proliferate (or conversely, to stagnate) without the gathering of fairly definite evidence, either by the developers themselves or by some public agency, as to the character and extent of possible harmful effects." The NAS committee also warned that "society simply cannot afford to assume that the harmful consequences of prevalent technological trends will be negligible or will prove readily correctable when they appear."[31]

In the case of the biotechnology industry, the concern about possible biohazards appears to have diverted the attention of scientists and policymakers during the late 1970s from the need and the opportunity to make such public assessments while the commercial applications were rapidly developing. Federal efforts were initially focused primarily on human health risks. Studies on the broader issues were initiated late and are of limited scale and scope. The Environmental Protection Agency has contracted for a study to produce an assessment of "the potential ecologic, economic and social impact" of the applied genetics industry, which is expected to be completed in 1982. The Office of Technology Assessment study, completed in January 1981, covers a number of related issues. A Congressional Research Service report on biotechnology prepared for the Subcommittee on Science, Research, and Technology of the House Committee on Science and Technology has recently become available.[32] It provides a useful overview which could help stimulate needed public discusision of the aims, directions, priorities and potential social and ethical impacts of the development of biotechnology.

Although there are ethical dimensions of all of the issues discussed thus far, several ethical problems related to genetic research and its applications have been of special interest. Recombinant DNA, along with other new techniques such as rapid gene sequencing, cell fusion and mass tissue culture methods, may be applied to higher organisms, including humans. There is public concern about the ethical aspects of human genetic screening, amniocentesis and, more recently, in vitro fertilization and gene therapy. The potential long-term effects of applied genetics on the environment and on evolution have also been discussed in terms of ethical responsibilities. As a result of the Supreme Court patent decision, additional concern about the ethical implications of private ownership of living organisms has been voiced by many individuals and groups, including the National Council of Churches.[33]

Recent accounts in the scientific and popular press have called attention to the ethical decisions university biologists are now facing because of possible conflicts of interest arising from their involvement with industry. Reports of demonstrated or alleged violations of the NIH guidelines by a few researchers have also highlighted the ethical problems encountered under a system of self-regulation where the principal investigator has primary responsibility for ensuring that safe experimental procedures are followed.[34]

The interest in these issues among the public and within the scientific community provides an opportunity for serious, positive discussion. Biologists have a good record of concern about the ethical aspects of their work. Many biologists recognize that their work touches on deep human values and has important effects on society. Because of their special knowledge, they can anticipate and identify possible problems related to their work at an early stage and participate with other groups to help make choices in accordance with publicly discussed ethical and value systems. Many of the leading genetic researchers have stated their awareness of the need to help initiate public discussion of such issues when the time seems appropriate. To establish and maintain public confidence in their credibility and social responsibility, scientists must be among the first to speak out. However, some scientists denigrate those who first warned against potential hazards of DNA research. In addition, several of the biologists who originally expressed concern have publicly recanted. Attitudes of this kind may discourage younger colleagues from exercising their responsibilities as scientists.

Biologists in the 1980s face issues that pose special problems for their own professional roles, for ethical standards and for their relationships to the public. A vigorous effort should be made to encourage working scientists to consider these problems. Studies are needed of the aspects of the life of science and the social system of science that encourage or inhibit a scientist to develop an awareness of the ethical dimensions of research and the related responsibility of the researcher. At the same time, we should urge scientists and nonscientists to explore these issues together, in an effort to restore communication and confidence.

Conclusion

New applications of molecular genetics are rapidly changing the relations between science, government and industry in a research field leading the search for new knowledge about fundamental life processes. Recombinant DNA is only one of several new techniques developed during the past decade that have enormously enhanced the scope and power of molecular genetics. Industrial and medical applications in this field are developing at a remarkably fast pace and will have increasingly important effects on the scientific community, the universities and the public. The problems generated by the stunning success of this basic research field must be addressed. Issues involving safety, ethical choices and social and economic impact are intertwined with problems relating to patterns of government support for basic research, the role of the universities and the social organization and value system of the scientific community. A main thrust of policy in this field should be to help define the roles and responsibilities of scientists and the public in efforts to anticipate and shape change, rather than merely to react to it.

Notes

1. Quoted in Terry Davidson, "DNA Meets the Public," *Somerville Journal* (15 January 1981), p. 1.
2. "Harvard Rejects Role in DNA Company," *Harvard Crimson* (18 November 1980), p. 1. The company subsequently was established as the Genetics Institute and sought to set up operations in Somerville, Mass. After the city's public hearing on the issue in January 1981, the firm withdrew its application and subsequently attempted to establish a laboratory in Boston, provoking public hearings there as well. The major scientific figure in the Genetics Institute is currently head of Harvard's department of biochemistry and molecular biology.
3. Parts of the historical summary in this section parallel the account presented in Charles Weiner, "Historical Perspectives on the Recombinant DNA Controversy," in *Recombinant DNA and Genetic Experimentation*, edited by Joan Morgan and W. J. Whelan (New York: Pergamon Press, 1979), pp. 281-87. Archival source materi-

als on the controversy over risks are available for study in the Recombinant DNA History Collection at the Institute Archives, Massachusetts Institute of Technology. For a description of the collection see Charles Weiner, "The Recombinant DNA Controversy: Archival and Oral History Resources," *Science, Technology and Human Values,* Vol. 4 (Winter 1979), pp. 17-19.

4. John Abelson, "A Revolution in Biology," *Science,* vol. 209 (19 September 1980), pp. 1319-21. This article introduces an entire issue of *Science* magazine devoted to technical reports on the impact of recombinant DNA techniques on fundamental knowledge of the structure, function, expression and regulation of genes.

5. Useful sources for growth of the field are the "NIH Registered Recombinant DNA Projects" monthly computer printouts compiled by NIH. Because of recent revisions in the NIH Guidelines, much of the recombinant DNA research supported by NIH is no longer registered and is not included in current reports.

6. Estimate provided by the Environmental Protection Agency.

7. The *Federal Register* and the *Recombinant DNA Technical Bulletin,* published by NIH, provide documentation of the changes in the guidelines, as do the minutes of the NIH Recombinant DNA Advisory Committee.

8. The meeting of IBC chairpersons was held on 24-25 November 1980 in Washington, D.C. An account is given in the *Pharmaceutical Manufacturers Association Newsletter* (1 December 1980), p. 5, and a full report has been prepared by NIH for the Recombinant DNA Advisory Committee.

9. U.S. Congress, Office of Technology Assessment, *Impacts of Applied Genetics: Micro-Organisms, Plants, and Animals* (Washington, D.C., 1981).

10. U.S. Senate, Committee on Commerce, Science, and Transportation, *Industrial Applications of Recombinant DNA Techniques,* hearings before the Subcommittee on Science, Technology, and Space, 20 May 1980.

11. Risk assessments that were particularly influential include the Falmouth Workshop in June 1977 (proceedings published as a special volume of *Journal of Infectious Disease,* vol. 13 [May 1978], Sherwood Gorbach, ed.); the Ascot Workshop in January 1978 (*Federal Register* [28 July 1978], Part 4, App. E); and the Rowe-Martin experiment (Mark A. Israel, Hardy W. Chan, Wallace P. Rowe and Malcolm A. Martin, "Molecular Cloning of Polyoma Virus DNA in *Escherichia coli:* Plasmid Vector System," *Science,* vol. 203 [2 March 1979], pp. 883-87; and Hardy W. Chan, Mark A. Israel, Claude F. Garon, Wallace P. Rowe and Malcolm A. Martin, "Molecular Cloning of Polyoma Virus DNA," *Science,* vol. 203 [2 March 1979], pp. 887-92).

12. Barbara Rosenberg and Lee Simon, "Recombinant DNA: Have Recent Experiments Assessed All the Risks," *Nature,* vol. 282 (20-27 December 1979), pp. 773-74; Stuart Newman, "Tumor Virus DNA Hazards No Longer Speculative," *Nature,* vol. 281 (20 September 1979), p. 176. Discussions of risk assessment by the RAC are summarized in the minutes of its meetings.

13. NIH, "Final Plan Program to Assess the Risks of Recombinant DNA Research," *Federal Register,* vol. 46 (10 June 1981), pp. 30722-78.

14. Statements at hearing of the Cambridge Biohazard Committee, 28 October 1980.

15. Proceedings of these hearings are available in National Institutes of Health,

Recombinant DNA Research: Documents Relating to "NIH Guidelines for Research Involving Recombinant DNA Molecules", vol. 1 (August 1976); vol. 2 (March 1978).

16. The guidelines now state that at least 20 percent of the members of the RAC shall be knowledgeable in applicable law, standards of professional conduct and practice, public attitudes, the environment, public health, occupational health or related fields.

17. Attendance at RAC meetings is recorded in the minutes of the committee.

18. Among the few detailed evaluations are Rae Goodell, "Public Involvement in the DNA Controversy: The Case of Cambridge, Massachusetts," *Science, Technology and Human Values,* vol. 4 (Spring 1979), pp. 36–43; and Sheldon Krimsky, "A Citizen Court in the Recombinant DNA Debate," *Bulletin of the Atomic Scientists* (October 1978), pp. 37–43.

19. Rae Goodell, "The Gene Craze," *Columbia Journalism Review,* vol. 19 (November-December 1980), pp. 41–45; Spyros Andreopoulos, "Gene Cloning by Press Conference," *New England Journal of Medicine,* vol. 302 (27 March 1980), pp. 743–46.

20. Stephen P. Strickland, *Politics, Science and Dread Disease: A Short History of U.S. Medical Research Policy* (Cambridge: Harvard University Press, 1972). See also Strickland, *Research and the Health of Americans* (Lexington, Mass.: D. C. Heath and Co., 1978).

21. Andreopoulos, "Gene Cloning by Press Conference."

22. Supreme Court of the United States, *Diamond, Commissioner of Patents and Trademarks* v. *Chakrabarty.* No. 79-136. Argued 17 March 1980, decided 16 June 1980.

23. Among the responses were several articles in the *Hastings Center Report* (October 1980): Lee Ehrman and Joe Grossfield, "What Is Natural, What Is Not?" pp. 10–11; Key Dismukes, "Life Is Patently Not Human-Made," pp. 11–12; and Harold P. Green, "Chakrabarty: Tempest in a Test Tube," pp. 12–13. See also Harold J. Morowitz, "Reducing Life to Physics," *New York Times* (23 June 1980), Op-Ed page.

24. Philip Siekevitz, "Of Patents and Prizes," *Trends in Biochemical Sciences,* vol. 5 (September 1980), pp. vi, viii.

25. "Memo Spells Out Technology Transfer Policies," *Harvard Gazette* (31 October 1980), pp. 4–5.

26. Letter to the editor from Richard C. Lewontin, *Harvard Crimson* (16 November 1980); "Profit – and Losses – at Harvard," editorial, *New York Times* (13 November 1980), p. A34.

27. *Harvard Crimson* (18 November 1980), p. 1.

28. Ibid.

29. Quoted in Eliot Marshall, "Will Biocommerce Ravage Biomedicine?" *Science,* vol. 210 (5 December 1980), p. 1103.

30. Letter from Joshua Lederberg to Senator Gaylord Nelson, 15 June 1978, quoted in Andreopoulos, "Gene Cloning by Press Conference."

31. U.S. House of Representatives, Committee on Science and Astronautics, *Technology: Processes of Assessment and Choice, Report of the National Academy of Sciences* (Washington, D.C.: July 1969), pp. 32, 34, 35. The NAS report was dis-

cussed in the context of recombinant DNA issues in 1978 by Susan Wright (see Selected Bibliography).

32. Congressional Research Service, *Genetic Engineering, Human Genetics, and Cell Biology: Evolution of Technological Issues, Biotechnology (Supplemental Report III)*. Prepared for the Subcommittee on Science, Research, and Technology of the Committee on Science and Technology, U.S. House of Representatives (August 1980).

33. Minutes, Meeting No. 4, 15–16 September 1980, President's Commission for the Study of Ethical Problems in Medicine and Biomedical and Behavioral Research; Richard Roblin, "Human Genetic Therapy: Outlook and Apprehensions," in *Health Handbook*, George K. Chacko, ed. (New York: Elsevier-North Holland Publishing, 1979), pp. 103–14; "Ethical Issues in the Biological Manipulation of Life," in *Faith and Science in an Unjust World: Report of the World Council of Churches' Conference on Faith, Science and the Future*, vol. 2, Paul Abrecht, ed. (Geneva: World Council of Churches, 1979), pp. 49–68.

34. A series of hearings on ethical problems in the biomedical sciences, scheduled for spring 1981 by the Subcommittee on Investigations and Oversight of the House Committee on Science and Technology, will focus on data falsification, commercial pressures on university scientists, social impact of genetic engineering and fetal research.

SELECTED BIBLIOGRAPHY

Recombinant DNA Controversy

Bareikis, Robert P., ed. *Science and the Public Interest: Recombinant DNA Research*. Bloomington: Poynter Center, 1978.

Cooke, Robert. *Improving on Nature: The Brave New World of Genetic Engineering*. New York: Quadrangle/New York Times Book Co., 1977.

Goodell, Rae S. "Public Involvement in the DNA Controversy: The Case of Cambridge, Massachusetts." *Science, Technology and Human Values*, vol. 4 (Spring 1979), pp. 36–43.

Goodfield, June. *Playing God: Genetic Engineering and the Manipulation of Life*. New York: Random House, 1977.

Grobstein, Clifford. *A Double Image of the Double Helix: The Recombinant-DNA Debate*. San Francisco: W. H. Freeman and Co., 1979.

Holton, Gerald, and Robert S. Morison, eds. *Limits of Scientific Inquiry*. New York: W. W. Norton and Co., 1979.

Hutton, Richard. *Bio-Revolution: DNA and the Ethics of Man-Made Life*. New York: New American Library, 1978.

Jackson, David A., and Stephen P. Stich, eds. *The Recombinant DNA Debate*. Englewood Cliffs, N.J.: Prentice-Hall, 1979.

Krimsky, Sheldon. "Regulating Recombinant DNA Research." In *Controversy:*

Politics of Technical Decisions, edited by Dorothy Nelkin. Beverly Hills, Calif.: Sage Publications, 1979, pp. 227–53.

Lappé, Marc, and Robert S. Morison, eds. *Ethical and Scientific Issues Posed by Human Uses of Molecular Genetics. Annals of the New York Academy of Sciences,* vol. 265 (1976).

Lear, John. *Recombinant DNA: The Untold Story.* New York: Crown Publishers, 1978.

Morgan, Joan, and W. J. Whelan, eds. *Recombinant DNA and Genetic Experimentation.* New York: Pergamon Press, Ltd., 1979.

National Institutes of Health. *Recombinant DNA Research Documents Relating to "NIH Guidelines for Research Involving Recombinant DNA Molecules."* vol. 1 (August 1976), vol. 2 (March 1978), vol. 3 (September 1978), vol. 4 (December 1978) and vol. 5 (March 1980).

Research with Recombinant DNA. Washington, D.C.: National Academy of Sciences, 1977.

Richards, John, ed. *Recombinant DNA: Science, Ethics, and Politics.* New York: Academic Press, 1978.

Rifkin, Jeremy, and Ted Howard. *Who Should Play God?* New York: Delacorte Press, 1977.

Rogers, Michael. *Biohazard.* New York: Alfred A. Knopf, 1977.

Scott, W. A., and R. A. Werner, eds. *Molecular Cloning of Recombinant DNA.* New York: Academic Press, 1977.

Southern California Law Review/Symposium on Biotechnology and the Law: Recombinant DNA and the Control of Scientific Research, vol. 51, no. 6 (September 1978).

Swazey, Judith P., James R. Sorenson, and Cynthia B. Wong. "Risks and Benefits, Rights and Responsibilities: A History of the Recombinant DNA Research Controversy." In *Southern California Law Review/Symposium on Biotechnology and the Law: Recombinant DNA and the Control of Scientific Research,* vol. 51, no. 6 (September 1978), pp. 1019–78.

Wade, Nicholas. *The Ultimate Experiment: Man-Made Evolution.* New York: Walker and Co., 1977, rev. 1979.

Wright, Susan. "Molecular Politics in Great Britain and the United States: The Development of Policy for Recombinant DNA Technology." In *Southern California Law Review/Symposium on Biotechnology and the Law: Recombinant DNA and the Control of Scientific Research,* vol. 51, no. 6 (September 1978), pp. 1383–1434.

Industrial Applications of Recombinant DNA

Andreopoulos, Spyros. "Gene Cloning by Press Conference." *New England Journal of Medicine,* vol. 302, no. 13 (27 March 1980), pp. 743–46.

Boorstein, Robert O. "Harvard May Form Group to Develop Lab Findings." *Harvard Crimson,* vol. 171, no. 102 (29 July 1980), p. 1.

Bylinsky, Gene, "DNA Can Build Companies, Too." *Fortune*, 16 June 1980, pp. 144–53.

Chakrabarty, A. M. "Which Way Genetic Engineering?" *Industrial Research*, January 1976, pp. 45–50.

Czarnecki, Mark. "New Life for Sale: Mass-Produced Life Goes to Market." *MacLean's*, 16 June 1980, pp. 41–49.

Dickson, David. "How Safe Will Biobusiness Be?" *Nature*, vol. 283 (10 January 1980), pp. 126–27.

_____. "Inventorship Dispute Stalls DNA Patent Application." *Nature*, vol. 284 (3 April 1980), p. 388.

_____. "Patenting Living Organisms – How to Beat the Bug-Rustlers." *Nature*, vol. 283 (10 January 1980), pp. 128–29.

_____. "Registration Proposed for Private DNA Research." *Nature*, vol. 283 (24 January 1980), p. 323.

_____. "U.S. Drug Companies Push for Changes in Recombinant DNA Guidelines." *Nature*, vol. 278 (29 March 1979), pp. 385–86.

Elia, Charles J. "Genentech Inc.'s Initial Offering Will Provide Chance to Invest in Nascent Genetic Research." *Wall Street Journal*, 27 August 1980, p. 37.

Fox, Jeffrey. "Genetic Engineering Industry Emerges." *Chemical and Engineering News*, 17 March 1980, pp. 15–22.

Greenhouse, Linda. "Science May Patent New Forms of Life, Justices Rule, 5 to 4." *New York Times*, 17 June 1980, p. 1.

Hubbard, William. "The Industrial Potential of Recombinant DNA Technology, Some Specific Applications." *Vital Speeches of the Day*, vol. 45 (15 March 1980), pp. 342–47.

"Insulin Research Debate on DNA Guidelines." *New York Times*, 29 June 1979, p. A18.

Krimsky, Sheldon, and David Baltimore (dialogue). "The Ties That Bind or Benefit." *Nature*, vol. 283 (10 January 1980), pp. 130–31.

"New Hope for Going to Market with DNA." *Business Week*, 24 September 1979, pp. 64–68.

"New Row about Society in Genetic Engineering." *New Scientist*, 17 July 1980, p. 181.

Newmark, Peter. "Engineered *E. coli* Produce Interferon." *Nature*, vol. 283 (24 January 1980), p. 323.

"OSHA Looks at Occupational Regulations for Recombinant DNA Use." *Nature*, vol. 283 (7 February 1980), p. 512.

Parisi, Anthony. "Industry of Life: The Birth of the Gene Machine." *New York Times*, 29 June 1980, Section 3, pp. 1, 4.

Pharmaceutical Manufacturers Association. "Recombinant DNA Hearings: You've Come a Long Way, Baby." *PMA Newsletter*, vol. 22, no. 21 (26 May 1980), pp. 1–2.

_____."Supreme Court DNA Patent Decision Not Necessarily the Last Word." *PMA Newsletter*, vol. 22, no. 25 (23 June 1980), pp. 1–2.

Randal, Judith. "OK Patents on Man-Made Life Forms." *New York Daily News,* 17 June 1980, p. 1.

Saltus, Richard. "Bay Area Research Teams in Genetic Photo-Finish." *San Francisco Examiner*, 11 July 1979, p. 8.

Schmeck, Harold M. "Interferon's Structure Revealed in Study." *New York Times*, 3 June 1980, p. 63.

Wade, Nicholas. "The Business of Science." *Science*, vol. 207 (1 February 1980), p. 507.

————. "Cloning Gold Rush Turns Basic Biology into Big Business." *Science*, vol. 208 (16 May 1980), pp. 688-93.

————. "Recombinant DNA: Warming Up for Big Payoff." *Science*, vol. 206 (9 November 1979), pp. 663-65.

————. "New Horse May Lead Interferon Race." *Science*, vol. 208 (27 June 1980), p. 1441.

"Where Genetic Engineering Will Change Industry." *Business Week*, 22 October 1979, pp. 160-72.

Wilford, John Noble. "Human Growth Hormone Is Produced in Laboratory." *New York Times*, 17 July 1979, p. C1.

Williamson, Bob. "Biologists Get Their Genes Twisted." *New Scientist*, 20-27 December 1979, pp. 12-13.

"A Wonder Drug in the Making." *Business Week*, no. 2612 (19 November 1979), pp. 71-74.

Yanchinski, Stephanie. "Microbes at Work: Is the Price Too High?" *New Scientist*, 3 July 1980, p. 12.

————. "Patenting Life Is No Guarantee of Success." *New Scientist*, 26 June 1980, p. 373.

5
Choosing Our Pleasures and Our Poisons: Risk Assessment for the 1980s

William W. Lowrance

Introduction

It takes only a few highly charged terms to evoke the risk-assessment milieu of the past decade: DDT, the pill, saccharin, "Tris," asbestos, nuclear waste, Three Mile Island, smoking, black lung, Clean Air Act, Delaney clause, recombinant DNA, 2,4,5–T, "Reserve Mining versus EPA," Teton Dam, DC-10. . . .

This rash of accidents, disruptions and disputes has left the public and its leaders fearful that the world is awfully risky and that, although science can raise warnings, when crucial decisions have to be made, science backs away in uncertainty. Further, there is a feeling that as with medical catalepsy, in which the simultaneous firing of too many nerves draws the body into spasms, the body politic has been drawn into a kind of regulatory catalepsy by too many health scares, too many consumer warnings, too many environmental lawsuits, too many bans, too many reversals. A related complaint is that we are afflicted with excessive government intervention, often of a naive, or trifling or naysaying sort. Among professional analysts as well as members of the public, there is a conviction that many risk-reduction efforts are disproportionate to the relative social burden of the hazards.

Public apprehensiveness has a number of causes. Is life becoming riskier? Not in any simple sense. As the next section of this report will demonstrate, many classical scourges have been conquered; infants get a healthier start in life; on average people live longer lives than ever before. The historical record of floods, hurricanes, typhoons, tornadoes, earthquakes and other geophysical disasters shows a relatively constant pattern of occurrence over

William W. Lowrance is senior fellow and director, Life Sciences and Public Policy Program, Rockefeller University, New York, N.Y.

the centuries.[1] (It is worth noticing, however, that migration is setting more potential victims in the path of hurricanes in the Gulf states and on top of seismic faults in California.) What we are menaced by now are enormous increases in the physical and temporal scale and complexity of sociotechnical hazards. Of these, the most threatening are risks having low probability and high consequence, such as genetic disaster, nuclear war and global climate change. Too, alarm arises, in an almost paradoxical sense, because science has become so much better at detecting traces of chemicals and rare viruses and at identifying birth defects, diseases and mental stress. Often we know enough to worry but not enough to be able to ameliorate the threat. Warnings and accusations are amplified by the public media, often with unseemly haste. Worse, scientific hunches are announced as scientific fact, only to have to be withdrawn later. With all this, it would be surprising if the public's sensibilities were not battered.

Risk-related instabilities and confrontations afflicting industry and governance stem as much from problems of societal attitude and decision-making procedure as from deficiencies of technical analysis and performance. This essay will argue that assessment will be improved if hazards are characterized explicitly, so they can be faced; if risk-aversion efforts are oriented to agreed-upon societal goals; if comparative approaches are taken that provide perspective, reveal the relative effectiveness of programs and lead to generation of stable, defensible priorities; and if attempts are always made to weigh risks in appropriate context with benefits and costs. The paper will review some institutional efforts, problems of public perception, challenges to scientific integrity and authority and a list of new and underattended hazards. It will conclude with recommendations.

The Evolution of Mortal Afflictions

In his 1803 *Essay on Population* Thomas Malthus observed of Jenner's new vaccine: "I have not the slightest doubt that if the introduction of cowpox should extirpate the smallpox, we shall find . . . increased mortality of some other disease." This general expectation holds true today if, in addition to disease, we include noninfectious threats. The communicable diseases of smallpox, diphtheria, typhus, cholera, tuberculosis and polio have been conquered. So have scurvy, pellagra and other nutritional deficiency diseases. Infant mortality has dropped dramatically. As the toll from these causes has lessened, mortality has shifted toward degenerative diseases — notably heart disease and cancer — which are attributable either to personal life style or to causative agents in the environment. While the causes of death have changed, the average age of onset of fatal illness has moved higher. Life span has lengthened. Put crudely, we die now of stroke and cancer in part because we live long enough to do so.

Figure 5.1. Deaths from Selected Causes
as a Percentage of All Deaths, United States, 1900-1977.

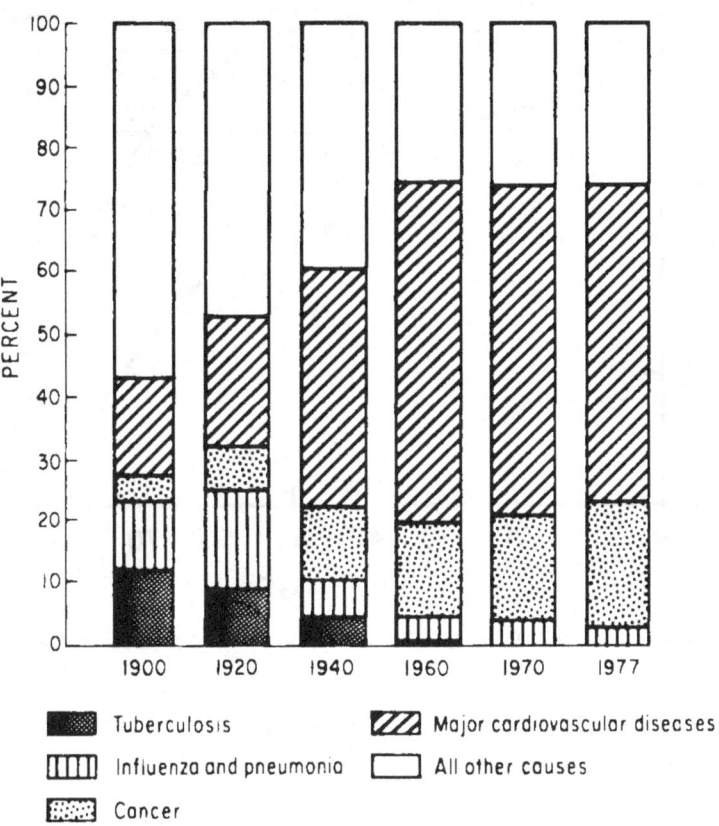

Source: U. S. Surgeon General, *Healthy People*, 1979.

For the United States these mortality trends are summarized in Figure 5.1.[2] Thus at present in the United States the leading cause of death is heart disease, followed by cancer. The rest of mortality is accounted for by other diseases and by accidents, homicide and natural disasters (in that order).[3] Within these gross statistics, however, there is great variability by age and socioeconomic status: Motor vehicles and other accidents kill the most children under 14; for black males between the ages of 15 and 24, homicide is the largest threat; cirrhosis of the liver is the fourth leading cause of death for people between 25 and 64.

In a recent analysis of the prospects for saving lives in this country, James Vaupel developed the concept of "early deaths." (The definitional problem

Figure 5.2. The Increasingly Rectangular Survival Curve.

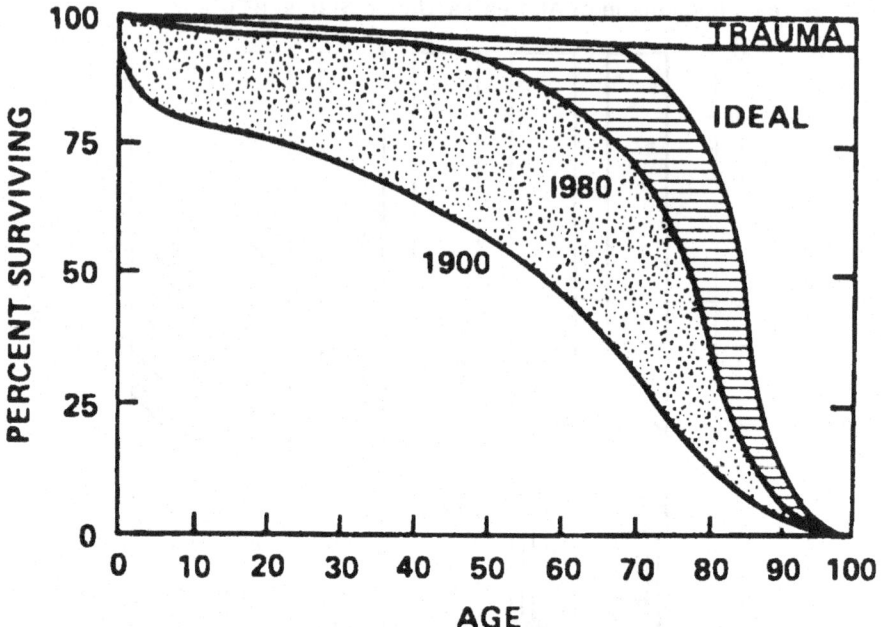

Source: James Fries, *New England Journal of Medicine*, vol. 303 (1980), pp. 130–135.

is fully treated in his report; for short, early death can be taken to refer to death before the age of 65.) Vaupel concluded:

> The statistics indicate that the aggregate social losses due to death are largely attributable to early death and that the losses due to early death are immense, that the early dead suffer an egregious inequality in life-chances compared with those who die in old age, and that non-whites, the poor, and males suffer disproportionately from early death. Furthermore, statistics on the leading causes of death and statistics comparing non-whites and whites, males and females, current mortality with mortality earlier in this country, and the United States with Sweden and other countries suggest that early deaths could be significantly decreased.[4]

Extrapolation of life expectancy data has led to another provocative observation about survival. Some analysts now speculate that the human species is approaching a "natural" life span limit of about 85 years. In Figure 5.2, the survival curve is seen to become increasingly "rectangular" and to approach a limit of 85 years. Such curves have led James Fries to predict that "the number of very old persons will not increase, that the average period of diminished physical vigor will decrease, that chronic disease will occupy a smaller proportion of the typical life span, and that the need for medical care in later life will decrease."[5] Even if the limit is inching upward, it is doing so at a low and decreasing rate; so the implications Fries draws should remain valid within the policy-relevant time frame.

Surely, coming to terms with these trends will lead us as a society to strive less to fend off full-lifetime mortality and to attend more to illness, accidents and quality of life. Among occupational diseases demanding attention, for instance, are the pneumoconioses: black lung disease, asbestosis and brown lung (textile dust) disease; among the most debilitating, lingering and painful conditions are arthritis, emphysema and allergies; among "life style" diseases, cirrhosis of the liver and the venereal diseases.

Improvements in Assessment

Becoming More Comparative

As a society we find ourselves, relative to all previous human confrontation with mortal risk, in the enviable but emotionally unsettling situation of living longer and healthier lives than ever before; of not having to remain ignorant and vaguely apprehensive of hazards but of understanding many of their causes, likelihoods and effects; and of having now accumulated substantial experience in predicting, assessing, reducing, buffering and redressing harm. Blissfulness is prevented by our having too many options. If we still lived only on the margin of survival, we would not have the luxury of worrying about microwaves and hairdriers. If we lacked scientific understanding and the prospect of taking preventative action, we would be more fatalistic about legionnaires' disease and toxic shock syndrome. If we had not established the hurricane warning network and the national air traffic control system, we would not have to argue about their budgets.

Howard Raiffa made the central analytical point recently in congressional hearings:

> We must not pay attention to those voices that say one life is just as precious as 100 lives, or that no amount of money is as important as saving one life. Numbers do count. Such rhetoric leads to emotional, irrational inefficiencies and when life is at stake we should be extremely careful lest we fail to save lives that could have easily been saved with the same resources, or lest we force our disadvantaged poor to spend money that they can ill afford in order to gain a measure of safety that they don't want in comparison to their other more pressing needs.[6]

To proceed in dealing with risks without making comparisons, both of import of threats and of marginal risk-reduction effectiveness (and cost-effectiveness) of public programs, makes little sense. Yet surprisingly little sophisticated comparative work has been done.

In studies meant to be illustrative, Bernard Cohen, Richard Wilson and others have assembled catalogues of common risks.[7] Cohen and Lee have calculated effects from different hazards upon life expectancy (for people at specified ages). They found that cigarette smoking reduces U.S. male life expectancy by six years on average. Being 30 percent overweight reduces life expectancy by about four years. Motor vehicle accidents cut off 207 days. And assuming that all U.S. electricity came from nuclear power and that the unoptimistic risk estimates published by the Union of Concerned Scientists are correct, nuclear accidents would claim 2 days from the life of an average :itizen. Regrettably, both Cohen's and Wilson's calculations are based on very unreliable data, fail to take into account indirect effects and are flawed in numerous ways. Their most valuable lesson has been to illustrate how difficult it is to reduce complex social phenomena, such as cigarette smoking and nuclear power generation, to single scalar risk rankings.

Stimulated in part by the early contributions of Chauncey Starr, over the last decade assessors have attempted to compare technological hazard to natural hazard.[8] For example, the so-called Rasmussen Report attempted to compare nuclear reactor accident risks to those of meteorite impacts and other natural hazards in order to provide some intuitive grounding.[9] The difficulty is that reliable numbers are hard to compute, and because polls have shown that most people, including scientists, do not have a very accurate intuitive sense of the likelihood and magnitude of natural hazards, such grounding may not be very useful anyway.[10]

The next logical step has been to try to compare the relative impacts various risk-reduction measures make on longevity. Shan Pou Tsai and colleagues, for example, have examined the question of what gains in life expectancy would result if certain major causes of death were partially eliminated. They calculated that for a newborn child, reduction of cardiovascular disease by 30 percent nationally would add 1.98 years to life expectancy at birth; 30 percent reduction of malignant cancers would add 0.71 years; and 30 percent reduction of motor vehicle accidents would add 0.21 years. If such 30 percent causative reduction were to exert effect during the working years of 15 to 60, there would be gains of 1.43 years (cardiovascular), 0.26 years (cancer), and 0.14 years (motor vehicle accidents). "Even with a scientific breakthrough in combatting these causes of death," the authors concluded, "it appears that future gains in life expectancies for the working ages will not be spectacular."[11]

Richard Schwing has published similar illustrative calculations of longevity extension. His findings for U.S. males, shown in Figure 5.3, chart the longevity increases and crude mortality rate decreases that would occur if certain causes of death were eliminated. It is obvious that further campaigns against tuberculosis would help few men and add only weeks of life for men on average, whereas reduction of heart disease would add years of life for a great many men.[12]

Schwing has gone on then, as others have, to compare the extent to which various risk-reduction measures—such as requiring that automobiles be built with energy-absorbing steering columns, penetration-resistant windshields or dual brake systems—extend longevity and to compare their cost-effectiveness (in dollars cost per person-year of life preserved).

Obviously the outcome of comparisons is heavily dependent on the way the boundaries of comparison are set. Nowhere has this been better illustrated than in recent attempts, such as the studies by the National Academy of Sciences (NAS) Committee on Nuclear and Alternative Energy Systems (CONAES) and by Herbert Inhaber, both of which compared competing energy cycles.[13,14] In calculating the risks of coal, do we count deaths from train wrecks, air pollution or release of radioactive radon from the burning fuel? In assessing nuclear power, do we include terrorist abuse or nuclear weapons proliferation? In appraising solar sources, do we include health effects on copper and glass workers? There is no avoiding such analyses. The problem is to learn how to perform them with technical sophistication and to take due account of all relevant social considerations. Overreaching is hard to avoid. The consolation of most such ambitious studies has been that the process of assessment has itself sharpened the social debate and clarified technical-analytic needs.

That the general public is sophisticated enough to understand and endorse the idea of comparative risk assessment has been demonstrated in such situations as Canvey Island in Britain. Within an area of 15 square miles on that island in the Thames near London are oil refineries, petroleum tanks, ammonia and hydrogen fluoride plants and a liquefied natural gas facility. When a few years ago controversy arose as to whether Canvey's 33,000 people were exposed to unusually high risks, a thorough government inquiry was conducted. Upon deliberation the residents passed a resolution that no further construction be accepted until the overall industrial accident risk on the island had been reduced to the average level for the United Kingdom. But they did not demand that their neighborhood be risk free.[15]

That the same toleration for comparative approaches holds in the United States is evident in industrial areas, such as Ohio and New Jersey, where residents are demanding cleanup, but not closing, of industries. Similar moderation led the voters of Maine, an environmentally sensitive state that has had to deal with cold winters but also with proposals of supertanker ports,

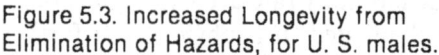

Figure 5.3. Increased Longevity from
Elimination of Hazards, for U. S. males.

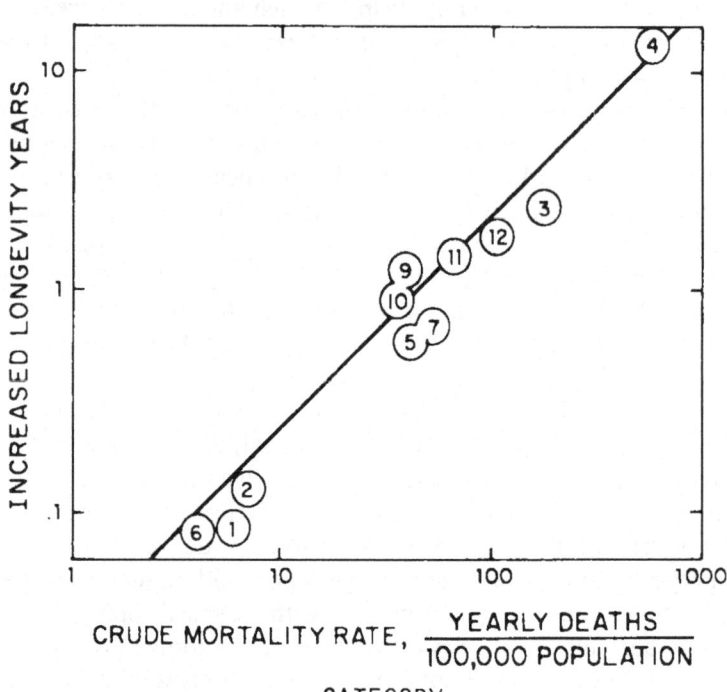

CRUDE MORTALITY RATE, $\dfrac{\text{YEARLY DEATHS}}{\text{100,000 POPULATION}}$

CATEGORY

(1) Respiratory tuberculosis

(2) Other infectious and
     parasitic diseases

(3) Malignant and benign
     neoplasms

(4) Cardiovascular diseases

(5) Influenza, pneumonia,
     bronchitis

(6) Diarrhea, gastritis,
     enteritis

(7) Certain degenerative
     diseases (nephritis,
     cirrhosis of liver,
     ulcers of stomach
     and duodenum, diabetes)

(9) Certain disease of
     infancy

(10) Motor vehicle
     accidents

(11) Other accidents and
     violence

(12) All other and unknown
     causes

Source: Richard Schwing, *Technological Forecasting and Social Change,*
vol. 13 (1979), pp. 333–345.

in their 1980 referendum to vote against measures that would have had the effect of being more restrictive of nuclear power.

If this country is to move toward more "rational" apportionment of risk-reduction and -management efforts, we must assure ourselves that there is reasonable parallel between the burden, in whatever terms, of particular risks and the avidity with which we defend against them, and that programs take into consideration age of onset of harm, degree of debilitation, longevity erosion and cost-effectiveness of ameliorative programs. Before any of this can be done, hazards have to be stated explicitly and goals of harzard reduction agreed upon.

Facing Hazards Explicitly

Comparative approaches are necessarily more quantitative, and they tend to force the revelation of specific consequences. As it dawns on social consciousness that even strict protection inevitably admits some residual harm, even if only by inducing exposure to the hazards of alternatives, little by little public officials have moved toward explicitness.

One of the most widely discussed test cases is that of DES (diethylstilbestrol, the growth hormone sometimes fed to beef cattle). The Food and Drug Administration (FDA) has formally proposed to allow beef producers to use this putatively carcinogenic but economically important agent, if they remove it from feed sufficiently in advance of slaughter that residual DES in marketed beef does not exceed a specified, extremely low concentration. In its proposal the FDA argued that "the acceptable risk level should (1) not significantly increase the human cancer risk and, (2) subject to that constraint, be as high as possible in order to permit the use of carcinogenic animal drugs and food additives as decreed by Congress. . . . A risk level of 1 in 1 million over a lifetime meets these criteria better than does any other that would differ significantly from it." The agency noted that further reduction "would not significantly increase human protection from cancer."[16] This proposal and similar ones are predicated on a conviction that the underlying carcinogen assessments are worst-possible-case overestimates of human risk. The DES standard is still under discussion. In March 1980 FDA Commissioner Jere E. Goyan stated that he would favor amending the food additives laws so that the chemicals testing out under the level of one chance in a million would be permitted (the Delaney clause prohibits even minute traces of very weakly testing carcinogenic additives – a prohibition honored mostly in the breach, because of its absolutist nature).

One by one, as cases have developed – the 1979 Pinto lawsuit, the national review of earthwork dams, amendment of the Clean Air Act – there has been a tendency to require that an upper bound on the estimated actual hazard be stated.

Specifying Risk-Management Goals

Although industrial and legislative programs usually operate under guidelines mandating "reduction of harm" or "protection of consumers," the degree of reduction or protection is often not specified (except when absolute protection is called for, which, usually being impossible, simply amounts to defaulting). Goal ambiguities may remain even when program objectives are spelled out. Different goals may come into conflict: reducing use of asbestos insulation, in order to protect miners and insulation installers, may have the effect of increasing fire hazard in buildings; forbidding black airmen who are sickle-cell-trait carriers to serve as Air Force pilots, to avoid the possibility of their becoming functionally impaired under emergency oxygen loss, conflicts with equal opportunity goals.

Recognizing that better guidance must be developed for choosing among the many available, but costly, marginal improvements in technical safeguards, the Advisory Committee on Reactor Safeguards of the Nuclear Regulatory Commission (NRC) has urged the NRC to consider establishing "quantitative safety goals for overall safety of nuclear power reactors." These goals might specify, for instance, physical performance criteria ("leaking of 10 percent of noble gas inventory from reactor core into primary coolant no more than once in 200 reactor years") or limits on health risk ("no more than one accident death per 1000 megawatts of electricity generated"). The advisory committee recently published *An Approach to Quantitative Safety Goals for Nuclear Power Plants*, and the commission has set in motion a "plan for development and articulation of NRC safety objectives."[17] Goals in this case include far more than the goal of generating economically competitive electricity.

A recent RAND Corporation study for the Department of Energy (DOE), *Issues and Problems in Inferring a Level of Acceptable Risk*, lists types of risk-reduction goals that can be considered, such as minimization of maximum accident consequences, minimization of probability of most probable accident and so on. After describing ways in which goal choices can make a difference to programs, the report urges that "DOE and other agencies need to be self-aware in specifying risk-reduction goals, as well as in relating them to goals of other agencies and interested parties, and understanding their implications for the choice of energy alternatives."[18]

Skeptics may be tempted to dismiss this topic, saying that we in this country do not have a consensus on social goals. Rebuttal to that too-simple dismissal is evidenced, for example, in the way our medical X ray protection practices, which are the result of decades of reassessment and improvement by industry, medicine and government, pursue goals: minimization of probability of damage (by decrease in frequency of use of diagnostic X rays, compensated for by more sensitive films), minimization of potentially irreversible damage to the human gene pool (special protection of gonads) and minimization of threat to infants in utero (again, special protection). The typically American goal of helping disadvantaged citizens underlies special health programs for minority groups. The goal of preserving maximum consumer choice can be seen as a goal of food quality programs.

Setting goals is not impossible, but setting realistically attainable goals is not easy. It is imperative that programs be tailored to goals more precise than "protection of all Americans against all harm."

Weighing Risks in Context with Benefits and Costs

All decisions, indirectly or directly, rely on judgments of the sort Benjamin Franklin referred to as "prudential algebra." Under the Toxic Substances Control Act, the Environmental Protection Agency (EPA) must protect the public against "unreasonable risk of injury"; under the stationary-sources provisions of the Clean Air Act, it must ensure "an ample margin of safety"; under the Safe Drinking Water Act, it must protect the public "to the extent feasible . . . (taking costs into consideration)." "Unreasonable," "ample" and "feasible" are not defined in these laws. For the EPA the question is not whether analysis but what form of analysis, taking what considerations into account. For all such risk-reduction regimes, the day has passed when benefits and costs could be ignored.

Every segment of industry and government — food, energy, transportation — has to ask:

- Are there ways to take benefits and costs into consideration along with risks? Do existing policy and managerial rules allow consideration of all such factors? Should they?

- Which methodological approaches (cost-benefit analysis, decision theory, cost-effectiveness analysis, etc.) are appropriate?

- How should secondary, indirect and intangible effects be taken into consideration?

- Are formal, explicit, published analyses required to form the basis of decision, or should they be used as informational background only?

- What are the procedural rules by which definitions, analytic boundaries and conceptual assumptions are established?

- Should those reviewing a technological option be required to review the attributes of alternatives also?

After a decade of concentrating on the negative side of the ledger, society is now trying to learn how to measure benefits. The NAS 1977 study of ionizing radiation ("BEIR II") struggled with the issue of how to appraise the benefits of such applications as medical X rays.[19] Its 1979 food safety policy report analyzed the benefits of saccharin and of food-safety policies regarding mercury, nitrites and aflatoxin (in peanut butter)[20] and its 1980 report, *Regulating Pesticides*, described the methods available for estimating marginal gains in crop yield and benefit expected from a candidate pesticide.[21]

Several methods, usually referred to in shorthand as "risk-benefit" or "cost-benefit analysis," are available for constructing a balance sheet of desirable and undesirable attributes. Analysis is thus a problem of handicapping what will happen (the odds of a destructive flood, the probable incidence of a disease) and comparing quantities that are rarely expressible in common-denominator terms (social cost of lives shortened, benefits of production, risks of genetic mutation).

With a few well-defined projects, for which goals and constraints are agreed upon by the major affected parties, for which health and environmental risks, costs and benefits are well known and understood (not only in magnitude but in social distribution, over both the near and long term), risk-benefit accounting has proven itself useful. Under such rare circumstances of certainty, commonsensical estimates as well as more formal analyses derived from operations research are applicable. The latter tend to be favored by specialists, technical or otherwise, who have been given a specific task to accomplish (the Army Corps of Engineers has pioneered in their use). The occasional "successful" application of such techniques—and, one suspects, also the all-embracing ring of their title—tempts legislators, administrators, managers and judges to call for their use.

The griefs of analysis could fill a large set of books. Most reviews conclude that such approaches are very useful for structuring discussion but are less useful, or even subject to misuse, when granted formal, legalistic weight. In their *Primer for Policy Analysis*, Edith Stokey and Richard Zeckhauser warned that:

> Benefit-cost analysis is especially vulnerable to misapplication through carelessness, naivete, or outright deception. The techniques are potentially dangerous to the extent that they convey an aura of precision and objectivity. Logically they can be no more precise than the assumptions and valuations that they employ; frequently, through the compounding of errors, they may be less so. Deception is quite a different matter, involving submerged assumptions, unfairly chosen valuations, and purposeful misestimates. Bureaucratic agencies, for example, have powerful incentives to underestimate the costs of proposed projects. Any procedure for making policy choices, from divine guidance to computer algorithms, can be manipulated unfairly.[22]

These and other critics respond to their own complaint by acknowledging that "prudential algebra" of one form or another must be resorted to, nevertheless.

All analytic approaches have difficulty with scientific uncertainties, with fair and full description of societal problems, with predicting all possible consequences, with placing a "price" on human life and environmental goods, with taking into account intangibles and amenities in general and with assessing the social costs of opportunities precluded.[23] A lively theater for this ongoing debate has been the proceedings of the Occupational Safety and Health Administration (OSHA) on regulation of occupational carcinogens.[24]

A somewhat different approach, "cost-effectiveness analysis," considers a present situation and compares how effectively alternatives can achieve stated objectives: automobile seatbelts compared with other forms of passive restraint, or kidney transplants compared with dialysis. Under the stimulus of cost-control campaigns, analysts have developed ways of comparing the relative cost-effectiveness of competing medical screening techniques and of other medical technologies.[25] Recently the congressional Office of Technology Assessment published a useful report on *The Implications of Cost-Effectiveness Analysis on Medical Technology.*[26]

Concluding a review for the Administrative Conference, Michael Baram argued:

> In practice, regulatory uses of cost-benefit analyses stifle and obstruct the achievement of legislated health, safety, and environmental goals. . . . Further, to the extent that economic factors are permissible considerations under enabling statutes, agencies conduct cost-effectiveness analysis, which aids in determining the least costly means to designated goals, rather than cost-benefit analysis, which improperly determines regulatory ends as well as means.[27]

Currently NAS is preparing a report, *Costs of Environment-Related Health Effects: A Plan for Continuing Study*, that should describe ways of building a base for accounting, in effect, for the health-cost-effectiveness of environmental controls.

There are other risk decision models. Jeffrey Krischer has recently prepared a useful annotated bibliography of applications of decision analysis to health care.[28] One of the more fully developed, unorthodox approaches is the libertarian synthesis of ethics and efficacy proposed by Ronald Howard.[29]

A concluding note should be that formal analysis is still helpless to accommodate many major effects: the weapons-proliferation and terrorist risks of the spread of civilian nuclear power, the highly touted and ambipotent benefits and risks of recombinant DNA development, the opportunity costs from undue conservativeness in regulation of contraceptive and pharmaceutical development.

Defining "Negligible" and "Intolerable"
and Setting Priorities

A disturbing feature of the 1960s and 1970s was that as each sector of manufacturing, or municipal governance, or research or purchasing found itself having to confront risk problems, each had to develop its own approach and work through hearings, scientific studies, economic reviews, lawsuits and insurance disputes. The social learning process was, unavoidably, painful. So were the disruption and unpredictability caused by the lack of defensible priorities. Industries and agencies found themselves so distracted by disputes over sensational cases that they could hardly pursue their main tasks, even if their charter was to reduce major risks: Neither "major" nor "minor" had been defined. Expressed in a metaphor of the time, smoldering barnfires had to be neglected while brushfires were fought.

Chastening has been accomplished. Now the challenge is to develop ways of keeping priorities clear: to avoid frittering away worry-capital on very small hazards, to prohibit unbearably large hazards and to concentrate decision-making attention on problems that affect large numbers of people in important ways. This admonition may appear an obvious one, but our failure to protect appropriate priorities is just what has set us up for the regulatory "overload" and disproportionateness we now labor under.

This concern was expressed in the 1980 NAS report, *Regulating Pesticides:*

> A serious flaw in the current procedure is that those compounds that receive the most publicity or pressure-group attention may not necessarily be those that present the greatest public health or environmental hazards. The current procedure does not provide for a broad comparison of the hazards posed by the large number of registered pesticides. At the same time, outside pressures to regulate a specific compound rarely arise from careful evaluation of comparative risks of alternative pesticides. To the extent that external pressures are influential in determining the order in which the [Office of Pesticides Programs] evaluates compounds, the consequence may well be that considerable resources are devoted to regulation of minor, low-risk compounds while important high-risk ones remain unreviewed for periods longer than would otherwise be the case.[30]

The March 1979 report by NAS on food safety policy proposed that the FDA categorize foods as being of high, moderate or low risk and "apply severe and general constraints only to items involving the greatest, most frequent, and most certain dangers."[31]

Naturally, regulatory agencies do try to apply their most vigorous attention to the most important issues, but their problem is to set protectable priorities (ones that are buffered from sporadic undermining) so that all parties involved know the analytic and legal agenda and can allocate resources accordingly. OSHA has tried to do this with occupational carcinogens, as has EPA with chemicals regulated under the Toxic Substances Control Act. The new National Toxicology Program is taking over some of the priority-setting tasks and will try to rationalize them across agency lines. The Consumer Product Safety Commission bases its priorities in part on a "frequency-severity index" derived from a computerized sampling system of hospital emergency-room admissions.

"Intolerable" and "unacceptable" are being invested with real-world connotations, as are "negligible" and "insignificant." These boundary-setting adjectives gain meaning in two ways: as experts, insurers and others rank hazards in hierarchies by severity, incidence and overall social exposure (hazards at the top and bottom of lists thus becoming obvious candidates for prohibition or acceptance); and as public opinion, lawsuits and so on indicate endorsement of the ranking. This helps administrators and managers allocate attention to the difficult cases in the middle.

In the beef DES example described earlier, some parties are urging that real, but very small, low-dose risks to humans be considered "negligible." The same principle is being appealed to in a current legal dispute over the regulation of the common hair-dye ingredient 4-MMPD. Seven hair-coloring manufacturers have sued the FDA for requiring that products containing 4-MMPD bear a label warning that the compound "has been determined to cause cancer in laboratory animals" and "can penetrate the skin." The plaintiffs argue that this stigmatizes the products, that scientific proof of 4-MMPD's carcinogenicity is weak and that, even if the chemical is carcinogenic to animals, "the risk is truly minuscule when compared to other potential or proven carcinogens . . . estimated to expose the individual consumer to a far greater risk of cancer than hair dyes containing 4-MMPD." A federal district court has remanded the case to the FDA, instructing the regulators to determine whether the chemical presents "a generally recognized level of insignificant risk to human health."[32]

In a striking case recently, the FDA approved the hair-dye chemical lead acetate. While acknowledging that in high doses the material is carcinogenic to rodents, the agency concluded that human exposure is so small, especially relative to overall lead intake, as not to warrant prohibition.[33]

Risk ceilings also can be established. In this country and many others polychlorinated biphenyls (PCBs) have been banned from commerce be-

114

Figure 5.4. Frequency-Versus-Severity Profiles for Nuclear Reactor Accidents in the United States, Relative to (a) Natural Hazards, and (b) Man-Made Hazards. (These curves are typical of their type but are meant here simply to be illustrative; numbers are debatable.)

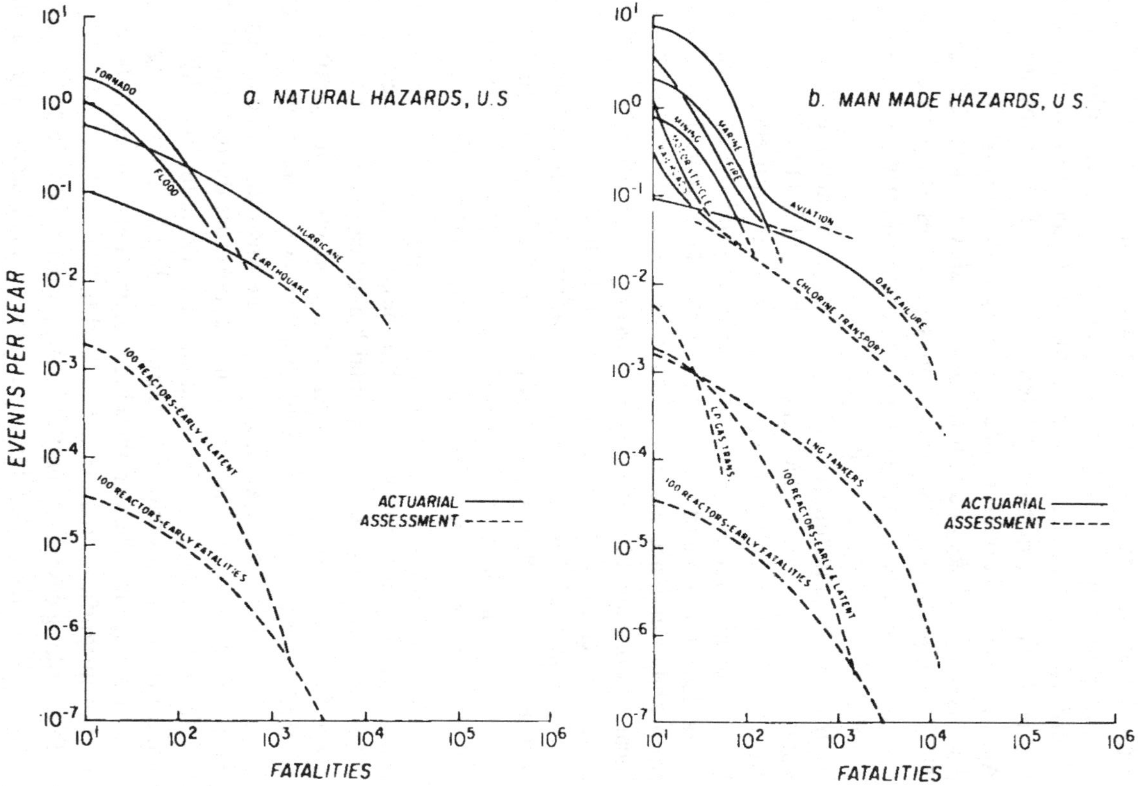

cause their carcinogenic potency is judged to be absolutely intolerable. From time to time, high-technology projects have been vetoed because their risks were unthinkably high: some macro-engineering modifications of the environment and certain potentially disastrous recombinant DNA experiments are landmark examples. The issue may not only be whether the hazards are actuarially high, but whether the threat would have an intolerably disruptive effect, physically or psychologically, on the fabric of society.

A useful way of envisioning risks is to profile them as a curve of frequency versus severity, as has been done for illustration in Figure 5.4. When a number of risks are plotted this way, certain domains can be recognized as de facto rejected (that is, society has repeatedly abjured risks in that range) or, on various grounds, as having been defined to be unacceptable. (The particular curves drawn in Figure 5.4 are typical of those under discussion currently; their numerical values are debatable. The basic method, though, of portraying cumulative, integrated risks in this fashion deserves exploration.)

Seeking Accommodation Between Technical and
Lay Perceptions

It is evident that "the public" often views risks differently from the way technical analysts do. (Of course, consensus is also rare, even within relatively closed circles of experts.)

From what do these differences of opinion stem? First, science itself is, in effect, simply a matter of "voting"; the scientifically "true" is no more than what scientists endorse to be true. Empirical knowledge is developed systematically within the scientific community, subject to criteria of repeatability, controlled observation, statistical significance, openness and the other guides of western science. By itself, procedure guarantees nothing, though. Good science is science that "works": science that can predict with consistency and generality and accuracy what will happen in the physical and social world. The weighing of facts remains subjective; perfect objectivity is a myth.

And second, judgments of hazards involve consideration not only of "size" of risks—likelihood and magnitude—but also of social value.[34] This, of course, leaves much room for disagreement.

Researchers have speculated that people's opinions about risks depend on many biasing factors, such as voluntariness of exposure, frequency of occurrence, amenability to personal control, reversibility, immediacy, bizarreness, catastrophic nature and so on.[35]

Social scientists such as Paul Slovic, Baruch Fischhoff and Sarah Lichtenstein have used polling techniques to survey risk perceptions and risk-taking proclivities. What they find, to neither their surprise nor ours, is that people have different perceptual biases. This research has concluded that human beings' brains, whether expert or lay, get overloaded with risk information and have trouble comparing risks; that the media accentuate social reverberations in risk disputes; and that, in essence, people believe what they want to believe. Person-in-the-street interviews of technical people show them to be not much better than nontechnical people at guessing, for example, how many fatalities are incurred annually from tornadoes, contraceptives or lawnmowers.[36]

Many of these polling studies are open to criticism. They suffer from the usual shortcomings of questionnaire design and the generic weaknesses of polling. Often they ask about only a single hazard at a time, which, by failing to foster or force comparison and by allowing people to express self-contradictory views, provides little guidance for policymaking. They force people artificially to break down their views into components. And these studies are vulnerable to being assumed (not necessarily by their authors) to imply findings about "the public," when in fact most of them have dealt with only small population samples.

Amos Tversky and Daniel Kahneman have developed an approach called "prospect theory" to account for the empirical findings.[37] Now perhaps theory will guide design of more sophisticated polls.

In the risk-assessment domain, as in others, we are being forced to realize that "the public" is a very elusive construct. No one person or group of people fully represents, or is representative of, all of our citizenry; and the "organized public" remains small and keeps changing in composition and opinion. For this reason, and others, the notion of "public participation" lacks conceptual shape. To oppose closed bureaucratic proceedings is usually legitimate, but it is a lot harder to devise proceedings that are not only open to the affected polity but that encourage extensive "public" participation without just opening channels for special-interest lobbying. A recent Organization for Economic Cooperation and Development study of public participation, entitled *Technology on Trial*, concluded: "the general thrust of participatory demand would appear to be for a greater degree of public accountability; freer public access to technical information; more timely consultation on policy options; a more holistic approach to the assessment of impacts: all of which amounts, of course, to more direct public participation in the exercise of decision-making power."[38]

In recent years both governmental and nongovernmental bodies have been taking steps to seek accommodation between lay perceptions and technical-analytic ones.[39] Regulatory agencies have opened up their proceedings and have solicited public input. Professional organizations have explored perceptual issues: In 1979 the National Council on Radiation Protection and

Measurement held a symposium resulting in a volume entitled *Perceptions of Risk.*[40]

If an attitudinal bias emerges, it can be incorporated into standards. In recognition of the public's extraordinary concern about catastrophic potential (as opposed to diffuse chronic risks) of nuclear reactors, for example, industry and its regulators have incorporated "risk aversiveness," or disproportionate conservatism, into reactor safeguards.[41]

A 1980 report, *Approaches to Acceptable Risk*, commissioned by the Nuclear Regulatory Commission, provides a very constructive review of many of these decisional problems.[42]

It is worth surmising that what is under perceptual dispute in many cases is not only the hazard itself but the social "management" of it. Nowhere has this been more bluntly evidenced than in the overall conclusion of the President's Commission on the Accident at Three Mile Island: "To prevent nuclear accidents as serious as Three Mile Island, fundamental changes will be necessary in the organization, procedures, and practices and—above all—in the attitudes of the Nuclear Regulatory Commission and, to the extent that the institutions we investigated are typical, of the nuclear industry."[43] Too, one suspects that risk opinions often may in effect be proxies for more deeply seated opinions about corporate bigness, or bureaucratic inaction or erosion of personal control.

Institutional Attention

Congressional Actions

As though swatting at swarms of hazards on all sides, during the 1970s the Congress passed, inter alia, the Consumer Product Safety Act, the Fire Prevention and Control Act, the Occupational Safety and Health Act, the Federal Water Pollution Control Act, the Toxic Substances Control Act, the Mine Safety and Health Act, the (aircraft) Noise Control Act, the Federal Environmental Pesticide Act, the National Earthquake Hazards Reduction Act, the Medical Devices Amendment to the Food, Drug, and Cosmetic Act, the Safe Drinking Water Act, the Resource Conservation and Recovery Act and various Clean Air Act amendments. To ensure independence of control, Congress split off the Nuclear Regulatory Commission from the old Atomic Energy Commission. And it established the Environmental Protection Agency, the Occupational Safety and Health Administration, the Consumer Product Safety Commisison, the National Fire Prevention and Control Administration and the Federal Emergency Management Agency to administer all the new laws.

The effect of this legislative crusade has been to bring tens of thousands of hazards into regulatory frameworks of many kinds, based on science, medicine, engineering, law and economics, that were — and still are — inadequate bases for decision.

The Congress has chosen a variety of roles for itself in risk assessment. It has established the regulatory agencies and overseen their work. With some issues, such as automobile emissions, it has insisted on reviewing the scientific and economic evidence in detail and on itself setting primary standards. With others, such as the arcane questions of recombinant DNA research, it has held hearings to establish a record but has refrained from instituting strong control (see Chapter 4 by Charles Weiner). Occasionally, in response to constituent pressure or political opportunity, it has intervened precipitously in regulatory action, as it has repeatedly done with saccharin, directing the FDA to stay an action or requesting the NAS to conduct another study. In emergencies it has held high-level inquiries, as it did during the Three Mile Island accident.

Recently the Office of Technology Assessment, the General Accounting Office, and the Congressional Research Service have all gotten more involved in preparing risk-related reports for the Congress. Congressman Don Ritter and others have proposed mandating that cost-benefit analysis be used as the basis for regulatory action; response to this bill in hearings has been mixed.[44] Congressman William Wampler has, in HR-6521, proposed creation of a National Science Council within the Executive Office (lodged in the Office of Science and Technology Policy), which would be charged with adjudicating major scientific disputes over factual matters in regulatory decision making. Prompted by such flaps as that over the questionable studies of health risks at Love Canal, legislators are considering establishing guidelines for scientific peer review of assessments used in regulation. Congressional concern over risk issues remains high, but it tends to focus on individual hazards rather than on a comparative high-risk-reduction agenda, and it tends to favor regulation as its best instrument.

Administration Actions

Various Executive Branch sagas in risk decision making have been described elsewhere and will not be reviewed here. We should, however, notice several trends that go beyond the straightforward execution of regulatory mandates.

There is some movement toward interagency coordination of regulatory actions. The complexity of the administrative task is illustrated by the fact that the Interagency Review Group on Nuclear Waste Management had to be constituted from 14 major entities of government (the Departments of Commerce, Energy, Interior, State and Transportation; National Aeronautics and Space Administration, Arms Control and Disarmament Agency, Environmental Protection Agency, Office of Management and Budget,

Council on Environmental Quality, Office of Science and Technology Policy, Office of Domestic Affairs and Policy, National Security Council and Nuclear Regulatory Commission.[45] The Interagency Regulatory Liaison Group (Consumer Product Safety Commission, Environmental Protection Agency, Food and Drug Administration, Occupational Safety and Health Administration and Department of Agriculture) has developed coordinated guidelines on carcinogenicity assessment.[46] A National Toxicology Program has been established to serve the needs of a number of agencies.

Fundamental research in support of regulatory work may be improving: The National Institutes of Health have become more involved in such matters as development of reliable and practical screening tests for carcinogens; the National Science Foundation now sponsors risk-related policy studies; and the National Bureau of Standards conducts fire research for the benefit of many agencies. How to marshall such support effectively is still a challenge: The basic research agencies don't have specific mission mandates, and the regulatory agencies lack strong fundamental research capabilities.

As part of its attempt to control economic inflation resulting from overregulation, in March of 1978 the Carter administration promulgated its Executive Order 12044, which directed the regulatory agencies to take a number of steps to "rationalize" their actions and to evaluate the promise of nonregulatory alternatives. Most controversially, the order called for economic impact analyses of major regulatory actions. As a result, a layer of procedures and organizations, such as the Regulatory Analysis Review Group and the Regulatory Council, was superimposed on existing, congressionally mandated agency structures. Adjustments to these developments have been painful. In his enlightening report to the Administrative Conference on these developments, Michael Baram concluded tactfully:

> Obviously, regulatory reform is in a state of flux, as COWPS, CEA, OMB, OSTP, RARG, RC, the agencies and Congress act in response to the stimulus of Executive Order 12044. New controversies have arisen as to the conduct and use of regulatory analyses, the adequacy of the methodologies employed, and the timing and extent of Presidential involvement in agency decision-processes.[47]

In its attempts to provide correctives for economically damaging overregulation, the Reagan administration will have to decide whether such centralized review is appropriate or whether such considerations can be delegated, with guidelines, to the agencies.

Court Actions

Thousands of tort cases are heard every year. For the present review, what is important are the ongoing debates over the role of the courts and the landmark decisions handed down by the high courts.

One respected view of the role of the judiciary is that championed by Judge David Bazelon: "Courts cannot second-guess the decisions made by those who, by virtue of their expertise or their political accountability, have been entrusted with ultimate decisions. But courts can and have played a critical role in fostering the kind of dialogue and reflection that can improve the quality of those decisions."[48] Others disagree, believing that courts should be free to review the substantive evidence and logic of assessments and decisions. The extent of judicial intrusion into agency decision making will remain an issue.

Recent years have seen the courts interpreting legislative mandates (as to whether, for instance, regulation under the Clean Air Act must consider costs, or whether the FDA properly interpreted its mandate in banning laetrile) and refereeing territorial disputes between agencies.

A crucial issue that continues to work its way up to the Supreme Court relates to the imperative for cost-benefit analysis in regulatory decisions. The recent case of *Industrial Union Department, AFL-CIO versus American Petroleum Institute* sidestepped the issue of whether OSHA must, under its statutes, base its decisions – in this case, over whether to tighten occupational exposure limits for benzene from 10 parts per million to 1 part per million – on formal, explicit, published cost-benefit analyses, the issue that many observers hoped the court would address.[49] The justices have, however, agreed to hear an analogous case, on cotton dust. The legislative background from which the Supreme Court has to work does not provide much guidance.

Nongovernmental Actions

Several recent developments exemplify the increasingly collective initiatives being taken by nongovernmental bodies. An impressive contribution has been made by the Food Safety Council, a non-profit coalition of industrial, consumerist and other members, which has developed and published a thorough review of the technical problems associated with food risk assessment and made proposals that are now under consideration by regulatory and other bodies.[50] The American Industrial Health Council, a coalition of 140 companies and 80 trade associations, has developed concerted positions on regulatory issues and is now proposing structural and procedural reforms.[51] In the aftermath of the Three Mile Island accident, the country's electric utilities and nuclear industry pooled their interests and established a Nuclear Safety Analysis Center, associated with the Electric Power Research Institute, to serve as an industry-wide reactor performance clearinghouse. Some 35 major chemical firms have recently established the Chemical Industry Institute of Toxicology, a research center charged with performing state-of-the-art toxicological research and assessment of large-volume commodity chemicals (not proprietary products) for the benefit of the industry as a whole. The major U.S. automobile and truck companies have joined the EPA in establishing a Health Effects Institute to study the effects of motor vehicle pollution.[52]

It is not yet possible to evaluate the promise of these new institutions. They deserve watching because they typify efforts to develop techniques, procedures, databases and focal centers for risk assessment outside of government. The question will be whether the work they produce is of high technical quality, whether they develop reputations of integrity and whether government and the courts can effectively accommodate the work of these hybrid institutions as alternatives to direct regulation and government-sponsored assessment.

It might also be mentioned that a Society for Risk Analysis has been formed, which (through Plenum Press) will in 1981 begin publication of a journal, *Risk Analysis.*

Scientific Integrity and Authority

Serious criticism is currently being leveled at the manner and quality with which scientific analysis is brought to bear on public hazards. Not to be interpreted as disaffection with science per se, this dismay reflects confidence that science can indeed help assess these problems, if it is properly applied.

Proposals are gathering for establishment of central authority structures to which technical disputes can be appealed. For example, the New York governor's panel (chaired by Lewis Thomas) formed to review the Love Canal fiasco found that "only further questions and debates on scientific credibility have been the result" of the "inadequate research designs" and "inadequate intergovernmental coordination and cooperation in the design and implementation of health effects studies" at the dump; as a remedy it recommended establishment of a Scientific Advisory Panel responsible to the governor.[53] Editorials have appeared in *Science* and elsewhere calling for reincarnation of the President's Science Advisory Committee to referee such disputes. Congressional Bill HR-6521 proposed formation of a National Science Council within the Executive Office for high-level review of assessments.

In somewhat the same vein, the American Industrial Health Council has urged Congress to establish a Science Panel:

AIHC advocates that in the development of carcinogen and other federal chronic health control policies scientific determinations should be made separate from regulatory considerations and that such determinations, assessing the most probable human risk should be made by the best scientists available following a review of all relevant data. These determinations should be made by a Panel of eminent scientists located centrally somewhere within government or elsewhere as appropriate but separate from the regulatory agencies whose actions would be affected by the determinations.[54]

Two questions must be asked of such proposals: whether "scientific and technical determinations" can legitimately be separated from "political and social determinations" and whether centralization of authority assures higher quality science.

To the first the answer is probably, yes, to a considerable extent, as long as it is understood that the very process of defining the problem is subjective and that scientific assessments usually have to be conducted iteratively. For example, to view the problem of liquefied natural gas facilities as one of time-averaged risk is different from worrying about the potentially massive social disruption one large accident could cause. Complex issues, such as energy policy, have to go many rounds of assessment, criticism, redefinition and reassessment.

To the second question, the answer is that communal scientific assessments do tend to gain critical analytic strength and social legitimacy over assessments made by individuals alone, but that pluralism and variety within the scientific community should be encouraged: recruiting more skilled policy-analytic scientists and engineers in industry, government and other organizations; appointing able advisory panels to many different administrative, legislative and managerial bodies; upgrading assessment work in academies, professional societies and trade organizations; and so on. Pluralism remains an essential safeguard against narrowness. Centralization and consistency are not always good in themselves. Besides, high-level bodies will always be limited to handling only a few contentious issues at a time. What they can do is raise warning flags about hazardous situations, draw attention to suspect scientific studies and help set the national agenda of assessment.

One of the more encouraging developments of the last few years has been a willingness of technical people, acting as professional communities, to review major assessments. When the original "Rasmussen Report" on reactor safety was issued, for example, it was subjected to detailed critique by a panel of the American Physical Society, by an ad hoc review group (the "Lewis Panel") chartered by the Nuclear Regulatory Commission, by the Union of Concerned Scientists and by others. Currently the Society of Toxicology is reviewing the controversial "ED-01" effective-carcinogen-dose experiment performed by the National Center for Toxicological Research.

New and Underattended Hazards

New hazards will always be cropping up, and there is no need to develop a complete new apprehension list here. The author believes that the following four can hardly escape becoming matters of heated controversy in the near future. The "thought exercise" is this: How can social and technical attention most effectively be brought to focus on them?

1. Women's Occupational Health. As women increasingly move into the heavy industrial workplace, there are questions of: (a) whether our scientific and medical understanding of women's bodies under stress is sufficient; (b) whether existing occupational standards protect all women as well as all men "adequately"; (c) if the answer to either (a) or (b) is "no," whether any health-related discrimination should be applied between the sexes (or, indeed, between small people and large people, or between any other categories by which human beings differ from one another) in conducting research and instituting protection; and (d) what actions should be taken specifically.

Part of this issue has to do with reproductive health, both of pregnant workers and of the fetuses they carry. Legal suits that have centered around this issue have not yet provided much clarification. Because mutation can occur in sperm, too, men are not exempt from danger. The Interagency Regulatory Liaison Group recently announced that it is conducting a major review of reproductive toxicology.[55]

Reproductive effects are not the only ones at issue: heat susceptibility, hearing loss, skin irritation and musculoskeletal damage may well turn out to be different for women.[56]

2. Urban and Indoor Pollution Hazards. Sealing up indoor environments hermetically keeps cold and smog out, but it may keep indoor pollutants in. Infectious and allergenic agents can be transmitted through an office's ventilating system. The problems of flaking asbestos and old lead-based paint are still with us. In a September 1980 report entitled *Indoor Air Pollution*, the General Accounting Office raised the alarm about various gases — radon, the radioactive gas released slowly from rock building materials; carbon monoxide, from various combustion sources; formaldehyde, from insulation; and others — that tend to build up and be circulated in sealed, poorly ventilated houses, mobile homes, offices and schools.[57] There continue to be allegations that nonsmokers are exposed to significant air pollution burdens from other people's smoking.[58] Continued urbanization and the campaign to insulate and seal buildings in efforts to save energy can only exacerbate these risks.

3. Teenage Pregnancy. It is hard not to be struck dumb by this problem. As expressed by James Vaupel:

One area seems particularly important. It involves the complex of overlapping problems associated with teenage birth, illegitimacy, prematurity, low birth weight, low IQ, deficient pre-natal and infant care, and high mortality rates not only for those children in infancy but also later on in life and for the mothers. The number of teenage births is startling: nearly 600,000 infants were born in 1975 to teenage mothers, some 240,000 to mothers age 17 or younger.[59]

Surely these numbers speak for themselves. Can any health risk be larger? To wave these problems off to "social welfare" bureaus and not address them along with the major issues on the national risk-reduction agenda is to take a very narrow view.

4. Seismic Hazard. Earthquake experts continue to predict major shocks for the West Coast. Engineers warn that although high, modern buildings are earthquake resistant, considerable peril remains in older, lower buildings. Fire hazard accompanies earthquake hazard in inhabited areas. As an exercise, officials might ask themselves how they will defend their current actions after the Big One strikes. Many of the problems are technical-economic ones that lend themselves to comparative analysis, as a recent Executive Branch review of California seismic hazard preparedness has argued.[60]

Recommendations

1. The overall urging of this essay is that bodies responsible for appraising public risk ask of their assessment efforts:

- Are risks, benefits and costs characterized as explicitly as possible?

- Are uncertainties and intangibles acknowledged and, where possible, estimated?

- Are programs oriented to agreed-upon societal goals?

- Do procedures guarantee that high-quality technical evidence is made available and used as the basis for decision?

- Are risks examined in a properly comparative context along with benefits and costs?

- Are precautions taken to prevent minor hazards from displacing larger ones on the protection agenda?

- Are the formality and legal bindingness of the analytic base appropriate?

2. Excerpts of well-regarded risk-assessment studies should be collected and published with commentary. (The NAS food safety study published several examples, and the NAS current review of some of its past projects — the "Kates study" — will provide more.) Critique should be made not

only of analytic methodology but also of how boundaries of assessment were set, how assessors were chosen, how conflicts-of-interest and biases were dealt with, how findings were expressed and how the study groups maintained their relationships with patrons and clients.

3. The causal connection between environment and health deserves continued investigation. As part of this, baseline surveys like the "LaLonde Report" (*Health of Canadians*) or the 1980 *California Health Plan* should be developed for the United States; this would be an extension of the 1979 *Report of the U.S. Surgeon General on Health Promotion and Disease Prevention.*[61] Then those determinants of health that are amenable to environmental influence should be evaluated.

4. The Office of Management and Budget, the Congressional Budget Office or others might direct or commission comparative evaluations of the marginal longevity gains and other benefits from key regulatory programs.

5. Evaluation should be made of such longstanding risk-management regimes as food inspection programs, fire-prevention provisions of building codes, flood plains insurance, black lung insurance and the like, asking whether they accomplish their risk-spreading or risk-reduction goals.

6. As the nation contemplates deregulation, sectoral net-assessment of regulatory policies should be conducted and reviewed. Alternatives to regulation should be examined, especially hybrid nongovernmental-governmental approaches.[62] In this regard the experiences of other countries, such as Sweden's in food safety, should be reviewed.

7. High-level scientific leadership needs continual renewal. One function of an upgraded White House scientific advisory body should be to identify major risk issues needing attention (such as, for example, the underattended issues cited at the end of this paper). This body, or other groups, should consider setting up a watchdog commission like the United Kingdom's Advisory Committee on Major Hazards to lead in the anticipation and assessment of important, long-term hazards.

8. There are many specific research needs, ranging from toxicology to policy analysis. Broad topics deserving attention include:

- Evaluation of the overall predictive usefulness of the toxicological gauntlet through which chemical products now are required to be run.[63]

- Improvement of epidemiology as an analytic complement to toxicological testing and continued development of the necessary databases.

- Refinement and comparison of such analytic techniques as cost-benefit analysis, decision theory, cost-effectiveness analysis.

• Evaluation of the validity of fault-tree and event-tree analysis as applied to nuclear reactors and other engineered structures.[64]

• Investigation of ways in which human error (maintenance error, operation error, emergency-response error) can be taken into account in probabilistic assessment of technological systems.

Notes

1. Ian Burton, Robert W. Kates, and Gilbert F. White, *The Environment as Hazard* (New York: Oxford University Press, 1978).

2. U.S. Surgeon General, *Healthy People: The Surgeon General's Report on Health Promotion and Disease Prevention, 1979*, U.S. Department of Health, Education, and Welfare Publication No. 79-55071 (Washington, D.C., 1979).

3. National Safety Council, *Accident Facts* (National Safety Council, 425 North Michigan Avenue, Chicago, Ill. 60611, 1980).

4. James W. Vaupel, "The Prospects for Saving Lives: A Policy Analysis," printed as pp. 44-199 of the U.S. House of Representatives, Subcommittee on Science, Research and Technology of the Committee on Science and Technology, *Hearings on Comparative Risk Assessment*, Ninety-Sixth Congress, Second Session (14-15 May 1980).

5. James F. Fries, "Aging, Natural Death, and the Compression of Morbidity," *New England Journal of Medicine*, vol. 303 (1980), pp. 130-35.

6. U.S. House of Representatives, Committee on Science and Technology, Subcommittee on Science, Research and Technology, *Hearings on Comparative Risk Assessment*.

7. Bernard L. Cohen and I-Sing Lee, "A Catalog of Risks," *Health Physics*, vol. 36 (1979), pp. 707-22; Richard Wilson, "Analyzing the Daily Risks of Life," *Technology Review* (February 1979), pp. 41-46.

8. Chauncey Starr, Richard Rudman and Chris Whipple, "Philosophical Basis for Risk Analysis," *Annual Review of Energy*, vol. 1 (1976), pp. 629-62.

9. U.S. Nuclear Regulatory Commission, *Reactor Safety Study: An Assessment of Accident Risks in U.S. Commercial Nuclear Power Plants*, WASH 1400 (NUREG-75/014). Washington, D.C. October 1975.

10. Paul Slovic, Baruch Fischhoff and Sarah Lichtenstein, "Rating the Risks," *Environment*, vol. 21 (1979), pp. 14ff.

11. Shan Pou Tsai, Eun Sul Lee and Robert J. Hardy, "The Effect of Reduction in Leading Causes of Death: Potential Gains in Life Expectancy," *American Journal of Public Health*, vol. 68 (1978), pp. 966-71. See also Nathan Keyfitz, "What Difference Would It Make If Cancer Were Eliminated? An Examination of the Taeuber Paradox," *Demography*, vol. 14 (1977), pp. 411-18.

12. Richard D. Schwing, "Longevity Benefits and Costs of Reducing Various Risks," *Technological Forecasting and Social Change*, vol. 13 (1979), pp. 333-45.

13. National Academy of Sciences/National Research Council, Committee on

Nuclear and Alternative Energy Systems, *Energy in Transition, 1985-2010* (San Francisco: W. H. Freeman and Co., 1980).

14. Herbert Inhaber, "Risk of Energy Production" (AECB #1119) (Ottawa: Atomic Energy Control Board, 1978). A summary version was published in *Science*, vol. 203 (1979), p. 718. Letters of criticism were published in *Science*, vol. 204 (1979), by Rein Lemberg, p. 454; Richard Caputo, p. 454; and John P. Holdren, Kirk Smith and Gregory Morris, pp. 564-66.

15. U.K. Health and Safety Executive, *Canvey: An Investigation of Potential Hazards from Operations in the Canvey Island/Thurrock Area* (London: Her Majesty's Stationery Office, 1978).

16. U.S. Food and Drug Administration, "Chemical Compounds in Food-Producing Animals: Criteria and Procedures for Evaluating Assays for Carcinogenic Residues," *Federal Register*, vol. 44 (1979), pp. 17070-114.

17. U.S. Nuclear Regulatory Commission, *An Approach to Quantitative Safety Goals for Nuclear Power Plants*, NUREG-0739 (Washington, D.C.: October 1980), and *Plan for Developing a Safety Goal*, NUREG-0735 (Washington, D.C.: October 1980).

18. Steven L. Salem, Kenneth A. Solomon and Michael S. Yesley, *Issues and Problems in Inferring a Level of Acceptable Risk*, Report R-2561-DOE (Santa Monica, Calif.: RAND Corporation, 1980.

19. National Academy of Sciences/National Research Council, Advisory Committee on the Biological Effects of Ionizing Radiation, *Considerations of Health Benefit-Cost Analysis for Activities Involving Ionizing Radiation Exposure and Alternatives* (Washington, D.C.: National Academy of Sciences, 1977).

20. National Academy of Sciences/National Research Council, Committee for a Study on Saccharin and Food Safety Policy, *Food Safety Policy: Scientific and Societal Considerations* (Washington, D.C.: National Academy of Sciences, 1979).

21. National Academy of Sciences/National Research Council, Committee on Prototype Explicit Analyses for Pesticides, *Regulating Pesticides* (Washington, D.C.: National Academy of Sciences, 1980).

22. Edith Stokey and Richard Zeckhauser, *A Primer for Policy Analysis* (New York: W. W. Norton, 1978).

23. David Okrent and Chris Whipple, "An Approach to Societal Risk Acceptance Criteria and Risk Management," #UCLA-ENG-7746 (Los Angeles: UCLA School of Engineering and Applied Science, June 1977); Baruch Fischhoff, "Cost Benefit Analysis and the Art of Motorcycle Maintenance," *Policy Sciences*, vol. 8 (1977), pp. 177-202; Dan Litai, "A Risk Comparison Methodology for the Assessment of Acceptable Risk" (Ph.D. dissertation, Massachusetts Institute of Technology, 1980); David Okrent, "Comment on Social Risk," *Science*, vol. 208 (25 April 1980), pp. 372-75; Chauncey Starr and Chris Whipple, "Risks of Risk Decisions," *Science*, vol. 208 (6 June 1980), pp. 1114-19.

24. U.S. Occupational Safety and Health Administration, "Identification, Classification and Regulation of Potential Occupational Carcinogens," *Federal Register*, vol. 45 (1980), pp. 5001-296; and OSHA Docket No. H-090.

25. David Eddy, *Screening for Cancer: Theory, Analysis, and Design* (Englewood Cliffs, N.J.: Prentice-Hall, 1980); Donald M. Berwick, Shan Cretin and Emmett B.

Keeler, *Cholesterol, Children, and Heart Disease* (Oxford: Oxford University Press, 1980).

26. U.S. Congress, Office of Technology Assessment, *The Implications of Cost-Effectiveness Analysis on Medical Technology* (Washington, D.C., 1980).

27. Michael S. Baram, "Cost-Benefit Analysis: An Inadequate Basis for Health, Safety, and Environmental Regulatory Decision-Making," *Ecology Law Quarterly*, vol. 8 (1980), pp. 473-531.

28. Jeffrey P. Krischer, "An Annotated Bibliography of Decision Analytic Applications to Health Care," *Operations Research*, vol. 28 (1980), pp. 97-113.

29. Ronald A. Howard, "On Making Life and Death Decisions," in *Societal Risk Assessment: How Safe Is Safe Enough?* Richard C. Schwing and Walter A. Albers, Jr., eds. (New York: Plenum Press, 1980), pp. 89-106.

30. National Academy of Sciences/National Research Council, Committee on Prototype Explicit Analyses for Pesticides, *Regulating Pesticides*.

31. National Academy of Sciences/National Research Council, Committee for a Study on Saccharin and Food Safety Policy, *Food Safety Policy*.

32. *Carson Products et al.* v. *Department of Health and Human Services* (Court for the Southern District of Georgia, No. CV 480-71, decided September 1980).

33. U.S. Food and Drug Administration, "Lead Acetate: Listing as a Color Additive in Cosmetics that Color the Hair on the Scalp," *Federal Register*, vol. 45 (1980), pp. 72112-18. The decision was lauded in a *New York Times* editorial, 9 November 1980.

34. William W. Lowrance, *Of Acceptable Risk: Science and the Determination of Safety* (Los Altos, Calif.: William Kaufmann, Inc., 1978); and William W. Lowrance, "The Nature of Risk," in Schwing and Albers, *Societal Risk Assessment,* pp. 5-14.

35. Paul Slovic, Baruch Fischhoff and Sarah Lichtenstein, "Facts and Fears: Understanding Perceived Risk," in Schwing and Albers, *Societal Risk Assessment,* pp. 181-216; Charles Vlek and Pieter-Jan Stallen, "Rational and Personal Aspects of Risk," *Acta Psychologica,* vol. 45 (1980), pp. 273-300.

36. Baruch Fischhoff, Paul Slovic, Sarah Lichtenstein, Stephen Read and Barbara Combs, "How Safe Is Safe Enough? A Psychometric Study of Attitudes Toward Technological Risks and Benefits," *Policy Sciences,* vol. 9 (1978), pp. 127-52; "Labile Values: A Challenge for Risk Assessment," *Society, Technology and Risk Assessment,* Jobst Conrad, ed. (New York: Academic Press, 1980), pp. 57-66.

37. Amos Tversky and Daniel Kahneman, "Judgment Under Uncertainty: Heuristics and Biases," *Science,* vol. 185 (1974), pp. 1124-32; "Prospect Theory: An Analysis of Decision Under Risk," *Econometrica,* vol. 47 (1979), pp. 263-91.

38. Organization for Economic Cooperation and Development, *Technology on Trial: Public Participation in Decision-Making Related to Science and Technology* (Paris: OECD, 1979).

39. Nancy E. Abrams and Joel Primack, "Helping the Public Decide: the Case of Radioactive Waste Management," *Environment,* vol. 22 (April 1980), pp. 14ff.

40. *Perceptions of Risk* (Washington, D.C.: National Council on Radiation Protection and Measurement, 15 March 1980).

41. J. M. Griesmeyer, M. Simpson and D. Okrent, *The Use of Risk Aversion in*

Risk Acceptance Criteria, #UCLA-ENG-7970 (Los Angeles: UCLA School of Engineering and Applied Science, October 1979).

42. B. Fischhoff, S. Lichtenstein, P. Slovic, S. Derby and R. Keeney, *Approaches to Acceptable risk* NUREG-CR-1614 (Washington, D.C.: U.S. Nuclear Regulatory Commission, 1980); also published by Cambridge University Press (New York, 1981).

43. President's Commission on the Accident at Three Mile Island, *The Need for Change: The Legacy of TMI* (Washington, D.C.: Government Printing Office, 1979).

44. U.S. House of Representatives, Committee on Science and Technology, Subcommittee on Science, Research and Technology, *Hearings on Comparative Risk Assessment.*

45. U.S. Interagency Review Group on Nuclear Waste Management, *Report to the President* (Washington, D.C.: U.S. Department of Energy, 1979).

46. U.S. Interagency Regulatory Liaison Group, "Scientific Bases for Identification of Potential Carcinogens and Estimation of Their Risks," *Journal of the National Cancer Institute,* vol. 63 (1979), pp. 241-68.

47. Michael S. Baram, *Alternatives to Regulation for Managing Risks to Health, Safety, and Environment* (a report to the Ford Foundation from the Program on Government Regulation, Franklin Pierce Law Center, White Street, Concord, N.H., 1 September 1980).

48. David L. Bazelon, "Risk and Responsibility," *Science,* vol. 205 (1979), pp. 277-80.

49. *Industrial Union Department, AFL-CIO versus American Petroleum Institute* (U.S. Supreme Court, decided 2 July 1980). Described in R. Jeffrey Smith, "A Light Rein Falls on OSHA," *Science,* vol. 209 (1980), pp. 567-68.

50. *Proposed System for Food Safety Assessment: Final Report of the Scientific Committee of the Food Safety Council* (Washington, D.C.: Food Safety Council, 1980).

51. *AIHC Recommended Alternatives to OSHA's Generic Carcinogen Proposal* (Scarsdale, N.Y.: American Industrial Health Council, 24 February 1978, OSHA Docket No. H-090).

52. Philip Shabecoff, "Health Institute to Study Motor Vehicle Emissions," *New York Times,* 12 December 1980.

53. Governor of New York, Panel to Review Scientific Studies and the Development of Public Policy on Problems Resulting from Hazardous Waste, *Report* (8 October 1980).

54. "AIHC Proposal for a Science Panel" (Scarsdale, N.Y.: American Industrial Health Council, 26 March 1980).

55. U.S. Interagency Regulatory Liaison Group, "Announcement of Reproductive Toxicity Risk Assessment Task Group Outline of Work Plan and Request for Comments," *Federal Register,* vol. 45 (1980), pp. 63553-54.

56. Vilma R. Hunt, *Work and the Health of Women* (Boca Raton, Fla.: CRC Press, 1979).

57. U.S. General Accounting Office, *Indoor Air Pollution: An Emerging Health Problem,* #CED-80-111 (Washington, D.C.: GAO, 24 September 1980).

58. James L. Repace and Alfred H. Lowrey, "Indoor Air Pollution, Tobacco Smoke, and Public Health," *Science,* vol. 208 (1980), pp. 464-72.

59. Vaupel, "The Prospects for Saving Lives."

60. U.S. Federal Emergency Management Agency, *An Assessment of the Consequences and Preparations for a Catastrophic California Earthquake: Findings and Actions Taken* (Washington, D.C.: FEMA, November 1980).

61. Marc LaLonde, *A New Perspective on the Health of Canadians* (Ottawa: Health and Welfare Canada Working Document, 1975); Office of Statewide Health Planning and Development, *California State Health Plan 1980-85* (Sacramento, 1980); and U.S. Surgeon General, *Healthy People.*

62. Baram, *Alternatives to Regulation for Managing Risks.*

63. U.S. Congress, Office of Technology Assessment, *Assessment of Technologies for Determining Cancer Risk from the Environment* (Washington, D.C., 1981).

64. Issac Levi, "A Brief Sermon on Assessing Accident Risks in U.S. Commercial Nuclear Power Plants," *The Enterprise of Knowledge: An Essay on Knowledge, Credal Possibility, and Chance* (Cambridge: M.I.T. Press, 1980). Also, enlightening analyses were presented in staff reports to the President's Commission on Three Mile Island.

PART 2

Science, Technology and International Security

6
Science, Technology and International Security: A Synthesis

Eugene B. Skolnikoff

INTRODUCTION

In a world substantially altered in this century as a result of the products of research and development, and with the elements of security of most nations directly affected, government institutions and policy processes in the United States remain heavily domestic in orientation. Contrary to common assumption, this is at least as true for the scientific and technological enterprise as it is for any other.

Some of the most important issues and needs relevant to science, technology and international security are presented in the following pages and in the accompanying chapters. The parochial nature of U.S. national institutions, however, makes it peculiarly difficult to come to grips with some of these needs or to anticipate them in any orderly way. For many years this problem has plagued U.S. government attempts to deal with the international implications of research and development (R&D) and international science and technology. The problems and the dangers now become more pressing as scientific and technological competence in other nations becomes more formidable. New measures are needed, yet the issue of excessive domestic orientation is only rarely identified or directly confronted. Without some attempt to understand this issue, actions that focus on the specific needs discussed below are likely always to remain ad hoc and seldom equal to their tasks.

Eugene B. Skolnikoff is director of the Center for International Studies, Massachusetts Institute of Technology, Cambridge, Mass.

BACKGROUND

The results of science and technology have had dramatic effects on the restructuring of nations and of international affairs, particularly in the 35 years since World War II. Aircraft, satellite communications, health and sanitation measures, missiles, nuclear weapons, automated production, radio and television, agricultural mechanization and new crop strains all bear witness to the productivity of R&D and, in their effects, to the profound revolution in human affairs they have brought about or made possible. The pace of change, furthermore, shows no sign of slackening.

International affairs have been heavily influenced by the differential ability of nations to carry out and capitalize on the results of R&D. Two nations have emerged with military power and influence far greater than others, largely as a result of natural endowments and resource bases that have allowed massive exploitation of science and technology. The gradual decay of that dominance, especially in its economic dimension, is already a source of new international relationships and problems. The disparity among nations of the North and South in ability to acquire and exploit technology is also a major factor in their relative economic status and in their increasingly acerbic political relations.

Concurrently, the pace of industrialization of technological societies has greatly intensified the dependency relations among states, so that even the most advanced societies find themselves critically dependent on others for resources, information, capital, markets, food and even technology.

Traditional geopolitical factors have been altered or expanded by advances in science and technology to include, inter alia, size and number of long-range nuclear missiles, satellite communications and surveillance capability, competence of the educational system, fundamental change in the very significance of major conflict and, critically, R&D capacity.

The results of R&D have also given rise to new technologies of global scale, creating wholly new issues in international affairs, notably atomic energy and space exploration. Also a matter of worldwide concern are the side effects of technological development. The resultant changes have altered traditional international issues and created major new ones, such as transborder environmental concerns, stratospheric modification and ocean exploitation.

Not all of these changes in international affairs directly bear on security, but the web of interactions in a technological world makes it difficult, even misleading, to exclude, say, economic concerns of developing countries from the concept of international security. In fact, the broad issues of food, health, resources, energy and population are aspects as legitimately a part of security as are military issues.

Given these effects of science and technology on the international security of states, it is ironic that the support for science and technology is primarily a national endeavor, particularly in the United States. Policies for R&D are seen in a national perspective and come primarily from national governments. This means, however, that international or global needs are not likely to be adequately taken into consideration in a national decision process.

A natural result of the nation-state system is that decisions in all policy areas are usually made unilaterally within one nation. Moreover, the apparent worldwide intensification of nationalism in the face of economic difficulty, not least in the United States, further encourages unilateral decision making. The parochial nature of decisions concerning R&D, however, goes beyond normal constraints of nation-based decision making and funding. The decentralized nature of public funding for research means that it is predominantly considered within the context of mission agency budgets. Even for those agencies whose rationale has a basic foreign policy motivation (Department of Defense, Department of Energy), the actual decisions and choices are heavily influenced by domestic pressures and inputs. Some departments or agencies are in fact precluded by their legislative charters from committing resources for anything other than domestic problems. All are faced with a budget process, in both the Executive and Legislative branches, that discourages (or often denies) all departments except foreign policy agencies the right to allocate their own R&D funds for other than U.S.-defined problems.

In the private sector as well, research decisions are heavily conditioned by the U.S. market, with U.S. industry still primarily concerned with U.S. sales and only gradually adjusting to the growing share of exports in the economy.

The implications of this situation are evident throughout the discussion of specific issues below and deserve subsequent elaboration to suggest possible policy or institutional departures that could be undertaken.

Of course, not all issues are handicapped by this particular institutional limitation. What follows is a broader discussion of the issues in the interaction of science, technology and international security that are likely to be central questions over the next five years. Though the focus is on a five-year period, policies cannot sensibly be seen in that short time frame without taking into account long-term objectives. Where relevant, what are, in effect, assumptions about desirable futures will be spelled out. The final section will be concerned with some of the institutional and policy process questions raised by the specific issues.

KEY ISSUE AREAS

It is tempting to start with national security issues, which appear to be most directly related to the subject. But economic issues will probably receive policy priority in the next few years, with important consequences for international security. In addition, as significant as defense issues are, they tend to receive more concentrated attention. Hence, defense issues will be addressed later in this paper, without in any way denying the fundamental significance of science and technology to security issues and, particularly, to international stability.

Economic Issues

Competition and Cooperation Among Advanced Industrial Countries

It is not a novel observation that the most serious short-term problem of the United States and of other Western industrialized nations is, and will continue to be, coping with inflation in a largely stagnating economic situation. Unemployment rates are high in many countries (over 9 percent in the United Kingdom at the end of 1980), with inflation at the double-digit level for several. This relatively bleak economic outlook has many causes; analysis of them within the context of this paper would be inappropriate. However, not only do economic problems affect the international role of science and technology, but some measures that individual countries may take for economic purposes will affect the course of science and technology or limit the international flow of scientific and technological information.

Industrial Policy. It has become almost a fad to speak of the need in the United States for an industrial policy or for reindustrialization. Several aspects of reindustrialization are particularly relevant to R&D. One is the ability (legal, political and psychological) of the U.S. government to work cooperatively with individual companies or a consortium to support research designed to improve the international competitive position of U.S. industry. Antitrust considerations, among others, have deterred such joint activity in the past.

Two initiatives in the Carter administration have shown that at least some of the barriers can be overcome. The joint research programs on automobile engines, with a consortium of auto companies (Cooperative Automotive Research Program), and the cooperative program for ocean margin drilling, with a group of oil companies, have received the advance blessing of the Department of Justice. These initiatives are now in jeopardy or cancelled. The international economic payoffs of cooperation of this kind (and the costs of not easing the way) may justify reconsideration of this policy in the next several years. Whether or not the government is involved, the advantage to international competitiveness of allowing research cooperation among companies in the same industry may create new support for antitrust policy legislation. Clearly, such legislation would provoke major political controversy.

A related aspect of industrial policy is the tendency of the United States to apply to U.S. companies operating abroad the same rules and constraints that apply inside the country.[1] The essentially adversarial relation between government and industry in the United States, whatever its historical justification or merits in spurring competition, often serves to put U.S. companies abroad at a disadvantage in competing with companies directly supported and often subsidized by other governments. This is particularly relevant in high-technology industries, as companies in other countries are now able to compete as technological equals for the major new markets that will determine future economic strength. Obviously many complex and contentious factors will arise as this issue is addressed, but they *must* be discussed. The economic stakes are high.

The key determinant of the U.S. competitive technological position is, of course, the strength and innovativeness of its high-technology industries. Domestic science policy, including support for research, tax incentives, regulations, quality and adequacy of education and other elements, will crucially affect the economic scene in years to come. In addition, specific tax and other policies that bear directly on industry's decisions to carry out R&D either abroad or in the United States will require examination, although it should not be an automatic conclusion that overseas research by U.S. firms is necessarily against U.S. interest. Overseas research can contribute directly to U.S. R&D objectives, enhance the possibilities for large-scale cooperation (more on this below) and contribute to knowledge generally.

One of the greatest dangers of the current economic malaise in Western countries, coincident with serious competition from Third World countries and from industrialized countries (especially Japan), is the possibility of a rise in protectionism—to preserve dying or inefficient industries. These industries may be failing for any number of reasons: increased labor costs relative to other countries; changes in cost of other factors of production, particularly for energy and resources; lower productivity; lagging innovation; inadequate industrial organization, and others. The temptation to respond politically to worsening domestic unemployment and its ancillary effects by preserving and protecting inefficient industries is very great, especially when a certain amount of implicit or informal protectionism is practiced by most countries in one way or another (hidden subsidies and biased regulations, for example).

The economic costs of a protectionist spiral among industrialized countries and the consequent loss of incentives for innovation and support of R&D could be very great. In effect, protectionist measures are an alternative to R&D investment, at relatively low short-term cost and very high long-term cost: a poor bargain, but one likely to be proposed and actively sought by powerful forces in the near future.

One specific protection issue has emerged in recent years over the export of new technology which, it is argued, is tantamount to the export of U.S. jobs as that technology becomes the basis of new competing industries. The argument is that technology developed in the United States is sold to others at a price that does not adequately reflect the true costs or the broader effects on the United States of that sale. It is a disputed issue, not only with regard to the facts, but also whether this is a case in which the possible cure might be worse than the disease. For example, is the current government pressure to exclude foreign students and faculty from advanced integrated-circuit research facilities at universities a wise policy? This is an issue likely to be more visible in the future.

Finally, under the heading of industrial policy, the relationship between domestic regulatory policy to protect health and safety and a nation's international economic position must be included. Already under intense scrutiny, this subject is certain to be the focus of important debate in the next five years. The basic concern is that unequal regulations from country to country can result in substantially different costs of production, thereby changing each nation's competitive position. That claim is made now with regard to U.S. environmental and safety regulations that are presumed to have important effects on U.S. export potential. Equalizing regulations worldwide would be one way to manage the problem when it exists, but that would not always reflect different conditions in countries, different factors of production, or different values. Regulations can sometimes improve competitive position if the costs of compliance are higher in other countries competing in the same market. At times, regulations are simply a disguised trade barrier. Once again, the complexity of the situation does not allow simple judgments or generalizations. The positive current account balance of the United States in the last months of 1980, in the face of high energy costs and an improving U.S. dollar value, would seem to belie the negative-effects argument, but it is not known what the balance would have been in the absence of regulation. Moreover, the issue is usually cast not only in specific cost terms, but also with regard to the delays, uncertainties and bureaucratic constraints imposed on industry by what is seen as a burgeoning regulatory environment.

The Reagan administration has indicated its intention to address this issue directly. It is hoped that sound data and analysis will support any actions taken.

Cooperation. Scientific and technological cooperation among Western technologically advanced countries is not rare. When compared with the scale of investments in R&D and the common goals of Western countries, however, the number of cooperative projects, especially in technological development, is rather small. The explanations are obvious: difficulties encountered in organizing cooperation; concern over losing a competitive position; and, most important, the basically domestic orientation of most governments. Meshing of programs, objectives, budgets and people is much more complex than when carried out within one country.

Current economic needs and constraints may now put cooperation, especially technological cooperation, much higher on the agenda. Industrial countries are all in need of technological progress to meet their social, political and economic requirements, at the very time when the economic situation that created these requirements also serves to place severe budgetary constraints on national R&D expenditures.

Today's nearly equal competence in science and technology among countries also means that a given project is likely to benefit from larger application of resources. In some cases, participation by more than one country may be necessary to attain a critical size. The massive investments required in many fields of central and growing importance, especially energy, also make the possibilities of cooperation to reduce the drain on national budgets particularly attractive.

The difficulties and costs of cooperation cannot be ignored:

- inherent difficulties of meshing disparate bureaucracies;

- delays in reaching decisions among differing political and legal systems;

- complications of varying decision processes, priorities and competencies;

- cost of international bureaucracy;

- the danger of political inertia, which makes projects hard to start but even harder to stop;

- the possibility of drains on research budgets because of international commitments;

- the tendency to undertake, internationally, only low priority projects;

- the apparent conflict between cooperation and improving a nation's competitive position.

Successful cooperation also requires reliable partners. The record of the United States in modifying or abrogating agreements makes future agreements harder to reach. Most recently, the proposals to cancel the coal lique-faction development project with Japan and Germany and to withdraw from the International Institute of Applied Systems Analysis have damaged our reputation as reliable partners.

Difficulties are formidable but the potential benefits are also formidable. Successful examples of cooperation (airbus, International Energy Agency projects, coal liquefaction until this year) demonstrate it can be done. Greater willingness of the U.S. bureaucracy to look outside the United States and recognize the competence and knowledge available elsewhere, and the greater experience the bureaucracy would attain through making the effort, would be substantial additional benefits of accelerating the pace of international cooperation. The forms of cooperation (bilateral, trilateral, Organisation for Economic Co-operation and Development—OECD) all need to be examined for each case, although the OECD is the logical organi-zation in which to lay the groundwork and establish a design among Western countries. Increased attention to genuine international technologi-cal cooperation ought to be an important task of the 1980s.

North-South Science and Technology Issues

The differential ability to acquire and exploit technology is a major deter-minant of the strikingly different economic situations and prospects of na-tions of the North and South and one of the prime sources of the political disputes among them. Differences in technological capability, however, are potential levers for constructive assistance and cooperation. Can this nation grasp those opportunities, which play to its strongest suit—its technological strength?[2]

The fate of developing countries in economic, political and military terms in coming years will have a great deal to do with international political sta-bility and with the security of all nations, not the least the United States. It is reasonable to forecast that international turbulence will be centered in the developing world. That estimate is reflected in U.S. military and foreign policies. It is much less evident in official economic policies—the U.S. com-mitment to economic assistance is scandalously low relative to that of other industrialized countries. The various reasons for U.S. indifference and fre-quent opposition to foreign assistance cannot be usefully probed here. However, the central nature of technology in development does provide a focus for exploring how to maximize the U.S. role, whatever the aggregate scale of assistance, and for highlighting some of the particular issues within specific fields (such as agriculture and population) that need to be con-fronted.

Economic growth, political stability and a working economy in a developing country (with important effects on agricultural production, re-source availability, reduction in fertility and markets for U.S. goods) can all be advanced by external assistance from the United States. It is in our national self-interest to provide this assistance. This is not to deny that the more economically advanced a developing country becomes, the more competitive it is with the United States; nor is it to deny that political stability does not automatically follow growth, or that the political objectives of developing countries may differ from our own. But U.S. self-interest is better served by the steady advancement of developing countries than by lack of progress. Whether or not economic assistance to developing countries is high on the U.S. agenda at the moment, there is a substantial probability that it will be forced there through political or economic crises or national calamities such as widespread drought.

Technology Policy Toward Developing Countries. It is no longer necessary to justify the importance of technology in development. Technology is essential to management of the problems of agriculture, health, environment, industrialization, population, energy and most other aspects of a modernizing society and is recognized (sometimes overemphasized) in most developing countries to be essential. The United States, whatever its relative decline in technological leadership, is still the world's strongest technological nation, with a broad and flexible education and research establishment.

The technological capability of most developing countries is steadily improving. Nevertheless, most research is carried out in the developed countries either for military purposes or for the domestic problems of those countries. Perhaps no more than 5 percent of global R&D can be said to be devoted exclusively to problems of development. In a setting in which industrialized nations have such a stake in economic growth and elimination of poverty in the developing world, it makes little sense to devote so little scientific and technological effort to problems that are peculiarly those of developing countries.

Much of this R&D cannot and should not be done in industrialized countries, for practical as well as philosophical and political reasons. To be effective, to work on the right problems, to be sensitive to local needs and preferences, to produce solutions that fit and are likely to be adopted, to keep up with and adapt technology—all require R&D defined and carried out locally. In turn, this implies attention to the building of the scientific and technological infrastructure in developing countries.

This does not mean, however, that all research relevant to developing countries needs must be carried out locally. Many areas of basic research can more effectively be done in existing laboratories; many problems are generic and can be more quickly investigated in established laboratories with resources and skills already deployed; many technological problems require general solutions before locally adapted applications are possible. Perhaps most important is finding ways to elicit commitments from scientists and engineers in industrialized countries to work on problems of development in a sustained way that allows cumulative benefits and continuous attention. Long-term availability of financial resources is essential, not only to make such commitment possible but also to make it respectable in the eyes of disciplinary peers.

Transfer of existing technology to developing countries is no longer seen as an adequate alternative. Experience shows that such transfer, especially of public technologies of health and agriculture, is inefficient or inappropriate without adequate receptors to choose, adapt, finance and develop knowledge to fit local environments and needs. Technology requires adaptation to a unique social, economic and political, as well as technical, environment. Also, it tends to change that environment, often quite rapidly, so that mutual adaptation of technology and environment is a continuing and dynamic process.

Relations of developing countries with multinational corporations also require local capability. The bulk of industrial technology is transferred to developing countries through private investment by international firms. To work effectively with technologically advanced companies, without losing control of the resulting development or being exploited economically, presupposes the ability to set realistic objectives, negotiate contracts, weigh often esoteric choices and, in general, be fully aware of technology and economic options.

Thus, a significant and growing indigenous capability in developing countries is required. And it must embrace basic science as well as technology, for without the insight and self-confidence created by an indigenous scientific community, a developing country will lack the ability to control its own development. In short, what is required is greater allocation of research resources to development problems in advanced countries, especially in the United States, and the building and strengthening of indigenous capability in developing countries.

To date, the ability of the United States to help in either of these efforts has been seriously limited, because of the low level of resources allocated and because of the institutional and policy constraints that deter or prevent effective commitment of scientific and technological resources for other than domestic purposes. At present, essentially all research devoted to problems of developing countries must come from the foreign assistance budget, either spent directly by the Agency for International Development (AID) or through transfer to other U.S. government departments and agencies. With minor exceptions, departments and agencies are prohibited by their legislative charters or by the budget process from spending any of their own funds on objectives other than domestic ones. Thus, in an overall federal R&D budget well in excess of $35 billion, the total allocated for objectives directly related to developing countries is on the order of $100 million, or one-third of 1 percent.[3]

The result is not only very limited in terms of R&D output; it also means that the competence of the U.S. government's technical agencies is barely tapped on issues to which they could significantly contribute. When all funds come by transfer from other agencies, there is no incentive to build staff or agency commitment, to work on these issues with relevant congressional committees and university or industry constituents or even to know through experience how these groups can contribute.

The rationale for these legislative restrictions and for budget compartmentalization stems from the early history of the creation of cabinet departments and agencies and from natural management principles of tying program objectives tightly to appropriate funding sources. The trouble is that, as foreign and domestic issues have become more closely intertwined, corresponding reflection in the allocation of resources has not taken place. And the rigid budget compartmentalization does not take into account the often mixed purposes (combining technological and development assistance goals) of many possible programs.

The implications of these institutional restraints go further. Astonishingly, the United States has no governmental instrument for cooperation with other countries, unless that cooperation can be defined either as scientifically competitive with domestic research and development, or as foreign aid for the poorest of countries. Thus, the United States cannot respond to those developing countries that have graduated from the poorest status, the very countries with developing science and technology capabilities best able to make use of cooperation with the United States, although not yet able to compete at the scientific frontiers. These countries have the greatest interest in substantive cooperation (often without any transfer of dollars) and are in the best position to begin solving their own problems as well as assisting in attacking global problems.

In fact, in recent years, the United States has undertaken rather substantial efforts at developing bilateral science and technology cooperation with these countries. Those initiatives have had to be taken primarily at the White House level directly, with major problems of planning and implementation. And now, at least some bilateral agreements that already have been negotiated may be abandoned as a result of large, targeted budget reductions.

The opportunities to use U.S. strength in science and technology in cooperation with other countries to further U.S. objectives (political and economic as well as scientific) are likely to grow in the coming years. The absence of an adequate institution and policy process to plan and fund these programs, as well as engage the competence of the U.S. scientific enterprise, both governmental and private, will be an important issue that will have to be confronted. The Institute for Scientific and Technological Cooperation (ISTC), which was proposed by the administration in 1978 and authorized but not funded by Congress, was designed to correct some of these institutional and process deficiencies.

Food and Agriculture. Some issues within the context of North-South relations stand out in their importance and in the likelihood that they will or should be the focus of much greater attention in the next quinquennium in the United States. One of these is food and agriculture, because of their fundamental role in the development process and the great concern that increases in agricultural productivity will not keep pace with the growth of population that already includes several hundreds of millions of chronically malnourished people.[4] It is estimated that food production must increase at least 3–4 percent per year if significant improvement of nutritional standards is to occur by the end of the century.[5]

The United States has a unique role to play because of its unparalleled agricultural production, as well as its R&D capabilities. For the reasons cited earlier, however, much of the necessary R&D and experimentation must be carried out in the countries trying to improve their own agricultural enterprises. This implies building greater indigenous capabilities than now exist and also strengthening and expanding the enormously successful international agriculture research centers that have been primarily oriented to, and staffed by, developing countries. The recent moves to devote more of the resources of these centers to the applied problems of improving agriculture (low-cost technologies, water conservation, etc.) are much to be applauded. The international centers must not be seen as alternatives to individual country capacity but as necessary complements to allow some economies of scale, to focus resources on generic problems and to provide an essential psychological tie to a world community for a sometimes isolated scientist in a poor country.

The U.S. research community could play a substantial role, larger than is at present likely. One impediment is the budgetary process, cited earlier, that bars the Department of Agriculture from effectively committing its own funds for agricultural problems not seen as domestic.

Another is the organization of agricultural research in the United States that is essentially a state-based structure without the extensive tools for central planning or quality control. This makes it difficult to ensure the essential quality of the entire agricultural R&D effort, to build competence in areas of study not peculiar to the United States or to enable effective, planned connections to be established between developing countries and the United States on agricultural R&D on any satisfactory scale.

It is also important to note that improvement in agricultural productivity is not dependent solely on advances in traditional areas of agriculture. Water conservation, climate, energy, pest control, low-cost technology and the social sciences related to agricultural economics, innovation, application and distribution, are, inter alia, of equal importance. The agricultural research agenda must include those areas as well.

Population. Although world fertility has declined in recent years, projected growth remains high enough to predict serious problems of starvation, economic stagnation and political unrest.[6] The international system has only begun to feel the effects of forced or voluntary migration across borders, which is likely to become a major cause of international political instability in the future; in addition, there is the already evident internal instability that arises from urban migration, unemployment or underemployment, lack of adequate food and sanitation and serious health problems.

Science and technology cannot solve the population problem, but they can provide the necessary tools for public policy. In particular, more research is needed to provide low-cost contraceptive technologies (especially male contraceptives) and to increase our understanding of the social determinants of effective family-planning policy. Fertility decline is so closely related to other aspects of development, particularly health, food, sanitation, transportation and communications, that in a sense all technological research can contribute indirectly or directly to the population problem.

In population-related (and health-related) subjects, we find a special variant of the domestic orientation of U.S. institutions. Health and safety regulation of drugs in the United States is based on risk-benefit criteria keyed to the United States. Thus, proposed contraceptive drugs are evaluated for safety based on the risks of health side effects in the U.S. environment, when the risks and benefits are likely to be quite different in another country. In some cases, U.S. pharmaceutical companies are deterred from developing a drug at all, because the benefits of protecting against some diseases (schistosomiasis, for example) are so low in the United States that any risk of side effects would overwhelm potential benefits, yet in another country the benefits would greatly outweigh the risks.

The reverse side of the coin is the stringent testing regulations in the United States that have led some companies to test drugs for safety in other countries, in effect using their people as guinea pigs for the U.S. market.

Neither situation is tenable. Some means must be found of internationalizing drug evaluation, as it would not be appropriate to expect the Food and Drug Administration, for example, to institute its own criteria for evaluating drugs for foreign applications that would be different from criteria for U.S. application.

The general problem of encouraging greater commitment of U.S. scientific and technological attention, whether in government, industry or university, to population- and health-related issues should be an important issue in the near future.

Transborder Issues

A series of transborder and global science and technology issues will be important elements of the international security picture in the next five years, although the separation of these from "economic" issues is rather arbitrary. The importance of environmental, ocean, resource and energy issues will be largely in their economic and, ultimately, political effects, as is the case for those just discussed.

Resources and Energy

In the short term, the major issues related to security, resources and energy have to do with supply interruptions engendered by political action and, secondarily, the economic terms on which resources are made available to industrialized societies.[7]

A major political phenomenon of recent years is the assertion of the right of absolute sovereignty over natural resources. It is a natural concomitant of a nation-state system but has not before been sanctified as it is today. The growing dependence of industrialized societies on resources under the control of others, particularly developing countries, creates major dependency relations, many fraught with great uncertainty and danger for international stability.

The dangers come not only from the threat of supply disruption, or of sudden dramatic increases in the cost of the resources, but also from the second-order strains created among industrial countries whose disparate dependence on resources from abroad may lead to major and disruptive foreign policy differences. The much greater dependence of Japan and continental Europe than the United States on Middle East oil, or the differential dependence on South African resources, could lead to serious conflicts of interest over Middle East or African or Soviet policy.

Although the world is painfully conscious of the political restrictions oil-rich developing countries sometimes place on resources, these countries are not the only ones to do so. Canada and Australia both have restricted export of uranium ore on nonproliferation grounds, and the United States severely restricts export of enriched uranium on the basis of specific political considerations. Moreover, the United States embargoed soybean exports for a short time in 1974 to stabilize domestic prices, and it has embargoed the sale of grain and high technology to the Soviet Union in protest against the Afghanistan invasion. A cabinet member of the Reagan administration in his first public statement spoke of using U.S. food exports as a foreign policy "weapon" (later changed to "tool").[8]

These consequences of resource dependency and of unequal distribution are all political and economic in character. The issues arising in the near future will be concerned with distribution and availability but not with depletion. In the long term, the adequacy of resources will be determined by economic, not geologic, phenomena,[9] and there is no reason to doubt that the industrial system could cope with long-term changes in the price and availability of materials and energy.

Short-term vulnerabilities must be met with measures that are largely outside the realm of science and technology directly: stockpiling, political negotiations, pooling arrangements in time of crisis and so on. Conceivably, new R&D for resource exploration, or exploitation of deep seabed minerals, could change U.S. dependency on foreign resources, but this is unlikely in a five-year time-horizon.

In the longer term, science and technology have major roles to play in the development of substitutes; in expanding knowledge of resource exploration, recovery, processing and use; and more generally in contributing to innovation and productivity in the nation's industrial plant (both to improve efficiency of use of materials and fuels and to generate the export earnings necessary to pay for imports). The long lead times inherent in reaching these objectives mandate early commitment of R&D to these tasks.

The changing price and availability of materials and energy may change critically the comparative advantage of some U.S. industries. The adjustments necessary to allow the orderly decline of those industries will themselves set up serious political and economic strains.

The need for R&D in the resource area is coupled with an inadequate understanding, both in the United States and globally,[10] of certain areas: geologic deposition of minerals, the exploration process and the impact of the changing industrial structure in minerals on the flow of mineral supplies.[11]

These tasks will require reinvigoration of concerned government agencies, especially the Bureau of Mines and the Geological Survey, and may also require a new institutional means to develop an objective, credible database (technical and economic) for resource-related decisions. In addition, coordination of policymaking must be improved to avoid conflicting policies carried out by individual agencies that are not aware of the activities of other agencies.

Environment and Global Commons

Closely related to resource and energy issues are those involving transborder environmental questions and more general global issues of the environment: atmosphere, oceans and outer space.

Our national activities have effects beyond borders and, in some cases, on a global scale. Transborder pollution has already become an important issue in many areas of the world, with some progress in the last decade, particularly in melding environmental policies, in reaching international agreements or in dealing with the traditional problem of the global commons. The issues are likely to become more severe, however, and often will take on the cast of zero-sum games.

The worldwide recession and the rise in energy prices raise the indirect costs of coping with environmental degradation and make it more difficult politically to restrict activities whose harmful effects fall across the border. The standard problem of reflecting full costs in a production process is exacerbated when the externalities are felt outside a national economy. Issues associated with acid rain, water pollution, forest degradation and others will become more contentious internationally in the next decade.

The depressed economic situation will also lead to greater resistance to domestic environmental regulation if that is assumed to affect adversely the international competitive position of a nation's goods. As noted earlier, it is not always appropriate to call for common environmental standards in all nations, and, even when it is, it is not clear they can be successfully negotiated. Thus, the costs and bases for domestic environmental regulations are likely to be difficult issues because of their international implications.

Some long-term issues may become clearer in the next few years as research increases understanding of important global systems. In particular, CO_2 buildup and NO_x in the atmosphere may be better understood, along with their global economic implications and potential ways of controlling them. Unprecedented disputes could arise over such issues, with important changes in the status of individual nations, as some benefit — say, through improved agricultural conditions — and others are hurt — for example, if the costs of environmental controls fall more heavily on them. It is unlikely that these issues will come to a head in a few years, but the debate could be far advanced.

Exploitation of global commons, especially the oceans and outer space, is likely to proceed during the coming decade. The Law of the Sea negotiation, which proposed a new international institution responsible for overseeing the mining of the resources of the seabed, appeared to be almost completed, although the position of the United States is now in doubt. Many aspects of that institution would be novel, in particular the assigning of some of the benefits of mining to developing countries. The detailed questions of implementation would be left to the interim arrangements following the completion of the treaty and ultimately to the new authority. Some serious disputes are inevitable, with regard to the mining itself, the operation of the authority and the unprecedented provisions for transfer of technology in the draft treaty.[12] Certainly, if there is no treaty, a variety of ocean issues — navigation, fishing, oil exploration, research, as well as mining — may become the source of serious dispute.

In space applications, controversy may arise over geostationary orbit allocations, but more likely controversy will be over the international efforts to manage and control space technology systems such as LANDSAT. This earth resource surveillance system has been until now an experimental U.S. monopoly, but as it moves to operational status, many questions will become more pressing. Who owns the information in a world in which sovereignty of resources has been zealously asserted? Should the output be available to anyone who asks for it? What rights do nations have for unilateral surveillance of another country's resources? What are the security implications of the high resolution that will now be built into the system? Who should manage the system and determine its technical characteristics? What are the economic and political implications of greater knowledge of resource endowments, of more accurate annual predictions of agricultural production domestically and internationally? Undoubtedly, these issues will soon become more prominent on the international political agenda.

Interaction of National Technological Systems

Many national systems – aircraft, communications, weather observation, finance, banking, postal – are basically information systems that require interaction with counterparts in other nations. The explosive development of information technology systems has begun to cause serious strains and is likely to be an even larger cause of strain in the coming years.

Traditional differences between fields break down (for example, communications versus data flows, postal versus electronic mail, information versus banking), and the economic calculus of benefits and costs changes perceptibly. Controversies arise over privacy of information, access to information within nations, the role of central computer banks, the transnational nature of economies of scale and related issues. In the face of U.S. dominance of technology, other Western countries are wary of allowing unfettered development that undermines their competitive position; the Soviet Union and its allies worry because control of information is vital to their political system; the developing countries worry that the loss of control over information will threaten their independence.

The dynamic nature of the growth of this technology, and its base in the private sector in the United States, makes this a particularly difficult issue in which to anticipate implications, much less develop clear international policies and conduct negotiations. It is certain to appear significantly on the international agenda in the 1980s.

National Security

Science and technology have been central factors in the evolution of weapons and military systems in this century. They have altered drastically not only the nature and scale of hostilities but the very meaning of strategic war as an option to achieve national objectives. The strength and productivity of a nation's advanced technological community have become major elements in any geopolitical calculation. Massive support for security-related R&D has, in turn, changed science, technology and the university.

The application of science to national security shows no sign of abatement. In fact, a new round of major commitments to large-scale strategic systems is in the offing, turning the ratchet one more notch in a search for security that seems steadily receding into the future.

In the context of this paper, only a few general issues in this area can be briefly touched upon; clearly it is an enormous subject that is itself the subject of a large literature.[13]

One controversy concerns whether the constant search for more technologically advanced weapons systems in fact contributes to the nation's (or the world's) security. Whatever the views of the causes of the arms race between the Soviet Union and the United States, or the current state of relations between the superpowers, new weapons systems often make the arms balance more precarious, more vulnerable to preemptive action, rather than contributing to stability. This may continue, and perhaps worsen, as capabilities are pursued that threaten concealment of weapons systems, give greater premium to surprise and make it harder to know whether missiles contain one or many independent warheads. Developments in conventional weapons, moving rapidly, may also change the nature of "local" war, leading to greater instability among developing countries as one or another believes it has the capability for rapid strike and victory.

No simple solutions exist. It is easy in rhetoric to call, for example, for more attention to military and related systems that contribute to greater stability and less uncertainty and threat: adequate conventional ground forces; improved command, control and communications in a hair-trigger weapons environment; greater commitment to developing arms control agreements; more attention to "hot-line" communication capability; less emphasis on strategic weapons that pose a first-strike threat in favor of those with clear survivability; and others. Each has its ambiguities, however, and there is no agreement on what is required for security, or even for greater stability.

The fact of the matter is that science and technology are most likely to continue to alter military systems. The effects of these changes cannot always be anticipated. One of the objectives of arms control is to bring the situation under greater control; but even if one were optimistic about SALT (Strategic Arms Limitation Treaty) II, agreements of this sort deal only with existing or planned technology. They do not deal with the possibility of new weapons systems or unanticipated capabilities created by further research.

Our knowledge of "threat systems," the involvement of the scientific and technological community in strategic debates, the public perceptions of military and strategic affairs are all inadequate. The once substantial public role of scientists and engineers in strategic policy deliberations, for example, has been greatly reduced, and the public inputs to arms control and weapons debates have suffered. This is illustrated by the spectacle of the stagnation of the SALT II agreement in the U.S. Senate over essentially extraneous issues.

Some argue that the whole framework of the strategic debate has been rendered inadequate.[14] They call for emergence of a new paradigm, a new discipline of conflict studies, and assign the scientific community special responsibility in bringing this about. The argument of the inadequate framework of debate is persuasive, although the path for achieving a new paradigm is hard to discern in practical terms.

The scientific and engineering communities have special but more traditional responsibilities within the existing framework, particularly because of the esoteric technical aspects of the issues. The relative neglect of these responsibilities in recent years must be reversed. New programs, such as arms control fellowships in the National Academy of Sciences and a concomitant program of studies, are to be applauded, and similar initiatives in other scientific organizations are to be encouraged. In all these efforts, however, it is important to recognize that the issues themselves are never purely technical. Real participation involves a commitment to master the political, economic and related aspects, which will eventually determine the outcome.

The quality of debate needs to be improved in the public sector as well as in the scientific communities. Better information and greater resources, public and private, committed to the analytical area are badly needed. The momentum of a defense budget close to $200 billion requires open debate of the purposes, details and implications of that budget. In turn, more funding is required to produce information and analysis to make public debate possible. The congressional commission to study the establishment of a National Academy of Peace and Conflict Resolution presumably has the same goal.[15]

One aspect of the role of science and technology in weapons development is peculiarly troubling. Much of the initial development of ideas for new technology — ideas that may later be revolutionary in military terms — occurs in the laboratory at a very early stage, without military applications in mind and often without military funding. This dynamic of the research process leads to instability, both in weapons development and in the long-term viability of arms control agreements.

Little can be done about this now, although ultimately ways of bringing R&D within the scope of arms control agreements must be considered. One aspect, somewhat farther along the R&D chain, does deserve institutional attention, however.

Proposals for new weapons development are, in their early stages, often made at low levels in the bureaucracy, with relatively little R&D funding required. At these levels, choices tend to be made on strictly technical grounds, with little consideration of their ultimate effect on relevant arms control objectives. The situation is repeated at higher levels as well, so that it is not uncommon for the government to be faced with mature weapons designs creating major new foreign policy problems that might have been avoided or eased if some alternative technical options had been chosen instead.

It is very difficult to deal with this issue in the bureaucracy, because the organization of government serves to create bureaucracies with compartmentalized objectives and a few or negative incentives to introduce considerations for which they are not responsible. An attempt to introduce nonproliferation considerations into planning for nuclear reactor R&D, through participation of a Department of State representative in the setting of objectives in the Department of Energy, has apparently had some limited success and deserves evaluation.

In its most general formulation, this task can be stated as the need to include the evaluation of broader effects of the intended results of research in defense R&D planning and management. The objective is an important one and ought to be the focus of further experimentation.

Other aspects of science, technology and security are also troubling, some because of the effects on nonmilitary areas. The sharp increase in defense spending proposed by the administration will have important effects on the civilian sector, not only in the obvious impact on the budget. Engineers, already in short supply, will be siphoned off in larger numbers to the defense industry, exacerbating the shortage in consumer goods industries and likely worsening the nation's competitive position. Increased spending will also tend to stimulate even more the momentum of scientific and technological change applied to military hardware, because the level of R&D, and the ideas for new applications, will be fueled by the larger cadre of scientists and engineers.

The increase in defense spending may also affect the nation's universities, as they become concerned about the almost direct military application of basic research. Signs of that are already evident in cryptological applications of theoretical mathematics, which have led to a kind of voluntary censorship.[16]

Lastly, it must be noted that the Soviet Union has demonstrated its competence to engage the United States in a high-technology arms race. Its technology may not be as refined, but its greater commitment of resources to defense expenditures is presumed by many to be likely to give the Soviets an edge of some sort over the United States in the latter part of this decade.

Whether this prediction is accurate or not, its anticipation has already fueled a massive new U.S. defense increase. One can only observe that a continued search for strategic superiority over a determined opponent is the search for a chimera that can only distract from the real quest for security.

East-West Transfer of Technology

Another issue that is likely to be of considerable moment in the next five years is the concern over the transfer of technology to the Eastern bloc that could enhance the military capability of the Soviet Union and its allies.[17]

This is an issue with a history stemming from the advent of the cold war, given recent attention as a result of the embargo on high technology imposed in response to the Soviet invasion of Afghanistan. It is bedeviled by controversy between the United States and its North Atlantic Treaty Organization (NATO) allies over the costs and benefits of the policy, by uncertainty over the military relevance of some "dual use" technologies, by sharp differences of view within the U.S. government, by differences of philosophy over the value of denial in terms of its actual effects and by differences with industry over enforcement policy.

There is little question about the importance of embargoing specific advanced military technology. Moving from technology with direct military applications, however, quickly leads to gray areas, with uncertainty over military relevance, over availability from uncontrolled sources or even of whether denial is in Western interests. Should the West, for example, encourage the Soviet Union to improve its ability to explore and recover its vast oil deposits?

Many more specifically technological questions arise, however. How is technology actually transferred and adopted? What is the real potential of diverting a piece of hardware from a peaceful to a military application? And what actual difference would it make? Is reverse engineering of a piece of equipment possible? At what cost? On what time scale? How long will it take for a particular technology to be developed?

All too often, the debate over technology export controls is characterized not only by political naiveté, as though it is simple to control the movement of technological information, but also by lack of understanding of technological realities. The importance of the issue, and its potential for damaging the West politically and economically, will require effective integration of the scientific and technological aspects in the policy debates.

INSTITUTIONS AND POLICY PROCESS

Several themes run through the issue areas discussed above that bear directly on the institutional and process problems of the United States in relation to the international consequences and use of science and technology. The most common theme is that the international dimension of policy is inadequately reflected in government policymaking and that the formal institutions of government militate against more effective recognition of international issues. Although this observation may be valid for many of the responsibilities of government, the problem is particularly, and surprisingly, intense in science and technology matters. Other themes that emerge relate to the need for more effective integration of scientific and technological aspects in many policy areas, including more mechanisms for effective analysis and anticipation of future implications of science and technology, and the need for new national and international institutions. Some comments on each are in order.

International Dimension in Policy

The history, geography and rich resources of the United States all led naturally to a system in which domestic considerations dominated institutional form and political organization. Adaptation of the system to its new global role, and to its new dependency on others, has been slow and halting, notwithstanding the enormous sums of public money allocated for this adaptation. At the level of detailed decision making—budget decisions, negotiations with the Congress or with the Office of Management and Budget, setting technical objectives—the traditional pressures dominate.

One of the most significant ways in which this situation affects the involvement of science and technology with international matters has to do with developing countries. The amount of national resources devoted to R&D on development problems is pitifully small, yet the U.S. government lacks an effective instrument for cooperating with that large number of increasingly important nations neither poor enough to be eligible for direct assistance nor sufficiently advanced scientifically to be competitive with domestic research. A new institution—the Institute for Scientific and Technological Cooperation—was proposed in 1978, authorized in 1979 and ultimately left unfunded by the Congress. Something to serve the same functions, whatever the form, is required.

But the problem is not simply a new institution. The need is to tap more effectively the scientific and technological resources of the government housed in the functional departments and agencies and to enlist their R&D clients in the nation at large. A single new agency cannot accomplish that task alone, although it might provide the leadership for much larger changes. Rather, a means must be found for allowing departments and agencies to allocate resources directly for cooperation with other nations and to carry out R&D on problems that are not "American" problems, when such activities are in the national interest. At present, legal authorization or executive budget policy effectively prevents such allocation except under difficult arrangements, sometimes sub-rosa and almost always ad hoc.

The problem is not primarily legal, as Congress can change the relevant laws and has done so for some agencies. The problem is largely one of efficient budgetary management. The Office of Management and Budget argues, with considerable justification, that it is difficult to maintain discipline in a budget if fuzzy arguments of "foreign policy interest" have to be given weight in ranking proposed programs, or if budgets to serve development assistance objectives crop up in a score of federal agencies.

Yet, the answer must surely be more creative than simply to rule out such programs. One possibility, for example, would be to create a development budget that crosses departmental lines and forces a degree of budgetary discipline that cuts across agencies and agency budgets. Departments and agencies would be allowed, with congressional concurrence, to budget some of their own funds for R&D, but those projects would have to be compared not only with proposals within the department but also with proposals of other agencies. Similarly, for those proposed programs that have mixed foreign policy (other than development) and scientific objectives, a cross-agency evaluation of foreign policy could exert the necessary budget discipline. Although difficult to administer and subject to its own bureaucratic pitfalls (the temptation for playing budgetary games and the difficulty of ranking according to foreign policy criteria), this evaluation or something like it requires experimentation.

In another area, ways must be found domestically or internationally to deal with situations in which apparently domestic regulations directly impinge on other countries or significantly affect a country's international trade position. For some situations, the answer may have to be regulatory machinery within existing or new international organizations. With regard to trade regulation, more impetus will have to be given to the move to analyze the broader economic effects of proposed regulations before the regulations are approved.

International cooperation with advanced countries also deserves more emphasis in the changing climate of cost and relative competence in science and technology. But this change in emphasis will not occur naturally in the U.S. system, again because of the built-in focus on domestic problems and pressures. This problem of focus is exacerbated by the restrictions imposed by the Office of Management and Budget on foreign travel and by the suspicion in Congress that foreign travel by "domestic" agency personnel simply implies junkets.

The blurring of domestic and international affairs is real. Government at all levels must become aware of and adapt to their ineradicable intertwining. It is not a matter of simply creating an international office in an agency. All have such offices, which more often than not are weak and removed from the core of the agency's interests. Rather, it is a matter of infusing the whole government with policies, institutions and rhetoric to make possible a gradual change of attitude that conforms to today's and tomorrow's reality. The Congress must also be no small part of that change and ought to be forcing the Executive Branch to recognize what is needed.

Integration of Science and Technology in Policy

The problems of scientific and technological planning are particularly severe and pose major problems of governance in a technological age. There are many aspects: how to represent scientific and technological information and uncertainty adequately in the policy process; how to plan for effects of science and technology not only uncertain, but possibly seen too late to alter once the effects are in evidence; how to estimate risks and benefits that fall unequally within a society or internationally, with interested people and nations often not represented in the policy process; how to deal with issues in which the relevant information is under the monopoly of one segment of society, or of one government; and a host of other issues.

No single solution is adequate. Like all problems of governance, these problems are not solvable – all that is possible is amelioration or improvement. However, these are difficulties that directly involve understanding of science and technology. Thus they require not only greater participation of scientists and engineers but also more means for making credible analyses available to the public and ways of drawing the public into the debates. Participation alone, of course, is not enough. Scientists and engineers do not have, on the basis of their professional training, superior credentials for making policy decisions. They are no freer of bias than are other segments of society. Participation by the scientific and technological communities implies a commitment to understand the interaction between science and technology and the broader aspects of policy, and a commitment of time that makes such understanding possible. A technocratic approach to the making of policy is not an improvement over the present situation.

One of the effects of science and technology on both national and international affairs is to make the future much more relevant to the present than in earlier periods of human history. To an unprecedented degree, today's policy must be made in the light of future developments, particularly in science and technology themselves or in the side effects of increasingly technological societies. The importance of more efforts at credible, objective anticipation of the future is obvious.

International Organizations and Structure

The need for new international instruments, or for modifying existing ones, was mentioned briefly in a few subjects – drug regulation, ocean mining, space applications – but was not emphasized. The questions associated with international political machinery, particularly machinery designed to deal with requirements growing out of science and technology, are many and complex.

The products of science and technology increasingly create new issues and force traditional domestic issues into the international environment. Unfortunately, existing international organizations charged with dealing with those issues are often inadequate. Most global organizations are now politicized along North-South lines, and more efficient regional or smaller alternatives do not represent all interested parties. As representation in organizations broadens, technical efficiency tends to decrease.[18]

This situation is unlikely to reach a crisis point within a few years, but in it are the seeds of major confrontation. These seeds could mature quickly, if current budgetary reductions drastically reduce U.S. presence in international organizations. The adequacy of international political machinery is likely to be a fundamental question of international security. So many of the functions the world (and the United States) depends on — communications, transport, nuclear materials control, resource information, health, agriculture, ocean minerals, to say nothing of international financing and lending — will fall increasingly under the auspices of international organizations. Many of the issues involve developing countries, but others involve conflicts of interest among Western industrial countries, or East-West controversies.

It is not a matter of indifference whether the organizations exist or work. The functions they perform must be carried out in some way by an organization, or by a limited number of countries or by a country acting on its own. The ultimate character of the international system and the place of the United States in it may in large measure be determined by whether these international tasks are carried out through organizations with broad participation but so designed as to allow reasonable efficiency or by default are managed by efficient but limited groups of wealthy countries.

CONCLUSION

It may not be too far wrong to characterize this last issue, and all that have been touched on in this paper, as fundamental choices in the international system between efficiency and equity and between hegemony and consensus. Those are sufficient for any policy agenda.

Notes

1. Lester Thurow, "Let's Abolish Antitrust Laws," *New York Times* (18 October 1980).

2. See Chapter 7 by Charles Weiss, Jr., in this volume.

3. *Development Issues,* 1981 Annual Report of the Chairman of the Development Coordination Committee, U.S. International Development Cooperation Agency.

4. The article by Sylvan Wittwer (Chapter 8 in this volume) discusses this issue.

5. Ibid.

6. The article by Michael Teitelbaum (Chapter 9 in this volume) is devoted to the population problem in substantial detail.

7. The article by William Vogely (Chapter 10 in this volume) deals with resource issues in detail.

8. "Block Changes Words, Terming Food as a 'Tool' Instead of a Weapon," *New York Times* (27 December 1980), p. 9.

9. See Chapter 10.

10. Ibid.

11. Ibid.

12. Law of the Sea draft treaty, United States Delegation Report, "Resumed Ninth Session of the Third United Nations Conference on the Law of the Sea" (Geneva: 28 July–29 August 1980).

13. The article by Kenneth Boulding (Chapter 11 in this volume) is devoted to one aspect of the subject.

14. Ibid.

15. Ibid.

16. *Science,* vol. 211 (20 February 1981), p. 797.

17. U.S. Congress, Office of Technology Assessment, *Technology and East-West Trade* (Washington, D.C.: U.S. Government Printing Office, 1979).

18. See E. B. Skolnikoff, *The International Imperatives of Technology,* Research Series No. 16 (Berkeley: University of California, Institute of International Studies, 1972), for a more complete discussion of technology and international organizations.

7

U.S. Policy Toward Scientific and Technological Development in the Developing Countries: The Case for Mutual Benefit

Charles Weiss, Jr.

Introduction

As the United States enters the 1980s, its foreign policy objectives are being reexamined to fit a changed world and a new political climate. These objectives must adapt to the greatly increased economic strength of Europe and Japan, to soaring energy costs coupled with threats of the cutoff of energy supplies and to the growing military, financial and political power of the developing world. Pressures for economic nationalism in the form of barriers to imports, exports and the flow of labor and technology are intensifying throughout the globe.

U.S. technology, although still the envy of the world, no longer enjoys undisputed preeminence. Difficulties in the automobile industry are only the most dramatic manifestation of the deterioration of U.S. competitiveness in relatively labor-intensive manufactures. The latest reexaminations of our standing in the international marketplace reflect concern that the United States may even be losing its competitive edge in electronics, hitherto one of its greatest strengths.

The world faces a future quite different from that optimistically projected in earlier decades. International inflation has resisted the prescriptions of a substantial range of schools of economic thought. The federal government's *Global 2000* study reiterates the conclusion of previous world models that the beginning of the twenty-first century will probably see a 50

Charles Weiss, Jr., is science and technology adviser, World Bank, Washington, D.C. This paper reflects the views of the author only. He has written it in his personal capacity. It is not an expression of the policies of the World Bank.

percent increase in world population, a doubling of real food prices, further increases in the real price of energy, worldwide loss of forests and genetic heritage and substantial pressure on the world's water resources.[1] For the first time, in the words of John Fairbank, the nations of the world are in trouble together.[2]

The developing countries (also known as "less developed countries" or "LDCs") are increasingly important to the United States. They are a source of oil and minerals, a market for exports now more important than Europe and Japan and a source of immigration, both legal and illegal. That their political instability can produce serious geopolitical consequences is seen in the fact that they have been the locus of every war since World War II. This instability is expected to continue, as even relatively optimistic projections place the number of people in developing countries whose low incomes deny them the most elementary requirements of decent living at 600 million in the year 2000.[3]

U.S. relations with the developing world must take into account the extraordinary diversity of a group that includes rapidly industrializing countries like Brazil, Korea and Yugoslavia; oil exporters like Nigeria, Indonesia and the Persian Gulf states; poor but technologically advanced states like India and China; rapidly growing exporters of agricultural commodities like Malaysia and the Ivory Coast; and relatively undeveloped, resource-poor countries like those of the Himalayas and the Sahel.

Our relations with these developing countries must also reflect the extraordinary progress that they have, in fact, made. While technical assistance remains crucial, relations with these countries increasingly require collaboration to meet shared long-range objectives. Such collaboration should in the long run replace the benefactor-to-client relationship and further reflect the great diversity of LDC political systems, national goals and overall attitudes to the United States, which range from close friendship to deep hostility.[4] As a final complication, LDCs at all stages of economic development and covering almost the entire political spectrum have found it in their interest to agree on common diplomatic positions and to negotiate collectively in the United Nations and other international forums under the rubric of the so-called Group of 77.

The United States can thus find itself in conflict with developing countries, individually or collectively, on matters of great political importance. In addition, U.S. interests often call for cooperation with a particular country in one area even when there is sharp conflict in another. We buy oil from Libya and collaborate on fusion research with the USSR, to cite two obvious examples. Hence, cooperation with developing countries on shared goals should not be automatically subordinate to fluctuations in bilateral relations, the North-South dialogue or other foreign relations concerns.

In today's circumstances, the national interest of the United States, and hence the priorities of U.S. foreign policy toward developing countries, need to be redefined as going beyond concern for the world's poor to include measures to protect its own economic and material well-being. Science and technology are important factors in both of these priority areas and should play a more important role in foreign policy toward LDCs.

Science, Technology and Foreign Policy Toward Developing Countries

Science and technology are both critical dimensions of development and the underpinning of U.S. economic and political strength. As such, they are major components of U.S policy toward the developing world. Scientific and technological collaboration with LDCs is cheaper than resource transfer and generally involves less immediate political cost. Science and technology are also key elements of many global issues that can be addressed only by international cooperation in which developing countries must play an important part. Finally, scientific and technological collaboration with these countries is essential to the solution of a number of important problems of scientific research. For all these reasons, such collaboration is critical to the long-term objectives of U.S. foreign policy.

It is also a two-way street. It can no longer be assumed that the problems of development can be solved simply by the "transfer of proven solutions" from the industrialized countries. Such solutions can have serious unanticipated effects if they are imposed without consideration of the economic, social, cultural and environmental conditions specific to LDCs.

What is more, the United States has much to learn from developing countries in such fields as urban transport planning, where the Singapore experiment in area licensing has greatly decreased traffic congestion and pollution; low-cost public health delivery systems, where the Chinese experience with "barefoot doctors" is a model for the world; and the use of fuel ethanol as a substitute for gasoline, where the Brazilian experiment illustrates both the techno-economic feasibility of the system where there is a surplus of arable land and the many social problems brought about by major changes in agricultural land use.

Science and technology have already played substantial roles in U.S. policy toward LDCs. But these efforts fall far short of either meeting LDC needs or achieving U.S. goals.

There are often political reasons to engage in scientific and technological collaboration for which the subject of that cooperation is unimportant. Scientific and technological cooperation may in itself be a gesture to symbolize or to help bring about a quick improvement in relations with a particular country. It may be an investment in the long-term development of cooperative spirit and of a cadre of people from the two countries accustomed to working with each other, or it may be quid pro quo for a specific but substantively unrelated diplomatic or political concession. This paper, however, stresses those areas in which the substantive results of scientific and technological collaboration serve shared interests, over and above the simple existence of the cooperative effort itself.

These scientific and technological aspects of development and of foreign policy defy easy classification. In this paper we distinguish four inevitably overlapping categories: (1) areas of technological development that are chiefly of humanitarian interest, such as public health, sanitation, subsistence agriculture and energy sources for the poor; (2) areas of technological development in which the United States and LDCs share material long-term interests, such as food and energy; (3) areas of national technological development, based on general U.S. interest in the development of a particular country; and (4) areas in which the United States must balance its interest in LDC development with other conflicting interests. Each of these areas is discussed in turn.

Technologies of Humanitarian Interest

U.S. policy toward the developing countries has been strongly committed to helping meet the basic needs of poor people. We strongly endorse this policy as it applies to scientific and technological development. Scientific and technological research has already had a major impact on the lives of the world's poor. Food production research has helped develop improved varieties of staples like wheat and rice. Research is advancing on "poor man's crops" like sorghum, millet and cassava. Recent biomedical research promises major advances in the prevention and cure of parasitic diseases, which afflict hundreds of millions of poor people. And the development of low-cost techniques for rehydrating victims of cholera and severe diarrhea has already saved many lives in refugee camps and in the poorest LDCs.

Advances in health-related technology occasionally produce a curious combination of relief from human suffering and narrow economic benefit. The eradication of smallpox eliminated an ancient human scourge that had in recent years been almost entirely confined to LDCs. At a much more prosaic level, this historic achievement also saved the developed countries the sizable sums of money needed to vaccinate their own populations and to verify the vaccinations of immigrants and visitors. Smallpox could not have been eliminated without the invention of the technique of concentrating vaccination efforts in high-prevalence areas and the introduction of the bifurcated needle for quick vaccination of large groups of people.

Another example of an area where humanitarian and financial motivations coincide is in the prevention or cure of diarrheal diseases. These diseases are among mankind's major health problems. They kill millions of children in developing countries each year. At the same time, successful programs would be worth billions of dollars to such diverse interests as multinational corporations, national governments and the international travel industry, all of which send large numbers of people to countries for which they are immunologically ill prepared. Health research in these areas is thus directly in the practical self-interest of the United States, over and above its critical, but commercially unattractive, humanitarian import.

Despite advances, much remains to be done. For example, it will be financially and institutionally impossible to provide sanitation facilities to urban populations of LDCs by 1990 unless alternatives to water-borne sewerage are used.[5] Better low-cost equipment is also needed to satisfy the energy needs of the poor. For example, simple cookstoves, made of clay and sand and constructed by local artisans for $10–25, have been readily adopted by test groups of housewives in Guatemala and Upper Volta and are reported to reduce the consumption of firewood by 40 percent.[6] Support is needed for both governmental and nongovernmental efforts (such as those of informal "appropriate technology" groups who are frequently in close touch with the poor) to develop and apply such technology at the grass-roots level. Such groups can in some countries help to overcome the social and cultural distance between urban-based scientists and technologists, trained along Western lines, and the poor people in slums and villages.

Technological Collaboration Based on
Shared Long-Term Objectives

In certain areas of technological development, the United States and the developing countries share clearly defined long-term interests. Food and energy are again clearly preeminent, although from a different point of view.

Food. According to the best available estimates, a continuation of present trends through the 1980s will result in major food deficits throughout the developing world, which can be met only by greatly increased exports from North America. Such exports, while resulting in gains for U.S. farmers, can be achieved only through much higher production costs and, hence, higher consumer prices, due to increased investments in fertilizer, water and other inputs and the lower productivity of marginal land brought into production. Future increases in oil prices and competition for land between food and fuel production and between agriculture and urbanization are likely to raise production costs still further, as is soil erosion caused by the decreasing willingness of farmers to pay the short-term costs of conservation measures.

In other words, U.S. consumers will face soaring food prices if the LDCs do not make major gains in agricultural productivity. Moreover, as a practical matter, there is every reason to believe that U.S. agricultural exports will continue to increase (unless the land they require is diverted to the production of fuel ethanol for domestic use). Therefore, it is in the direct interest of the United States to help LDCs mobilize technology to increase their agricultural productivity.

As mentioned above, scientific and technological research has already made major contributions to food crop production technology in developing countries. High-yielding varieties of wheat and rice (together with an institutional structure for training, extension and research that made it possible to take advantage of these varieties, adapt them to local conditions and convince local farmers to use them) helped to transform India from a significant importer of food grains in the 1960s to a marginal exporter in the mid-1970s. This averted both a major economic and human disaster in India and a substantial drain on the world's food production.

In another area, scientific and technological research enabled tracking of the meteorological patterns that influence breeding, swarming and migration of the desert locust, and the satellite monitoring of climatic features favorable to its increase, greatly reducing the threat of the desert locust in the Middle East and east Africa. By contrast, failure to incorporate genes for rust resistance into the wheat varieties distributed to farmers in another large, poor country led directly to a crop failure and a foreign exchange crisis a few years ago.

Energy. As the United States will for some time to come be a net importer of fossil fuels, it is critically important that a diversity of suppliers be available and that production of fossil fuels for world markets be sufficient to meet demand. Since U.S. reserves are unlikely to increase beyond their current size, and since many U.S. allies are inescapably dependent on foreign oil, the discovery of oil anywhere in the world is an important objective of U.S. foreign policy. Energy conservation and the use of renewable forms of energy anywhere in the world are likewise strongly in U.S. interest — not only because they reduce world demand for fossil fuels but also because they reduce the amount of carbon dioxide that is being added to the atmosphere and affecting the world's climate.

Credible geological estimates indicate that much of the world's undiscovered oil, natural gas and other fuel minerals lies beneath LDCs. Therefore, serious exploration and exploitation of conventional energy resources in these countries works to the advantage of the U.S. public. In other words, the United States has a direct interest in seeing that the LDCs have the capability to carry out geological explorations, deal with multinational oil companies, develop energy plans and policies, promote energy conservation and choose, adapt, and, when necessary, create a renewable energy technology suited to local circumstances.

Moreover, to the extent that developing countries have become self-sufficient in energy through conservation and through development of indigenous supplies — both renewable and nonrenewable — they are less dependent on suppliers in a particular part of the world and less prone to political instabilities brought on by balance-of-payment deficits. To the extent that LDCs avoid "showpiece" development of nuclear power over and above techno-economically justified requirements, they will tend to lessen the incidence of terrorism and weapons-grade plutonium proliferation. Finally, given the inevitability of the spread of nuclear power technology to LDCs, the U.S. public has a vital interest in the ability of developing countries to operate nuclear power plants in ways that avoid both accidents and diversion of nuclear material.

Science and technology are already working to solve LDC energy problems. Brazil's ability to replace much of the gasoline fraction of imported petroleum by fuel ethanol at competitive prices is largely due to indigenous adaptive engineering in locally manufactured equipment for crushing, fermentation and distillation.

By contrast, another example from Brazil shows how underestimation by the United States of the importance of the place of indigenous technological development can lead to unfortunate consequences. The United States, in accordance with its policy of opposing nuclear proliferation, provoked an international incident by objecting to Brazil's purchase of German nuclear power technology which included reprocessing equipment. Once the political need to stand up to U.S. pressure had passed, the Brazilians realized that they had overestimated their needs for nuclear power, that the contract provisions for technology transfer were not as favorable as they had first thought and that the program was likely to tie up a disproportionate share of their engineering manpower. Brazilian energy policy is apparently being readjusted accordingly.

In addition to food and energy, there are a number of other areas where scientific and technological cooperation with LDCs meets definable, if less concrete, U.S. interests. These range from financial interests to scientific and environmental concerns to broad-based political issues.

Ecological and Cultural Heritage. Much of the world's irreplaceable ecological and cultural heritage lies in the LDCs. This heritage includes such diverse treasures as ancient monuments and works of art, ancient cities (some of which still teem with people), game parks, the habitats of such endangered species as the Bengal tiger and vast stretches of undisturbed tropical rain forest. These forests contain countless endangered species of inestimable economic potential that constitute a genetic treasure incalculably greater than that of the celebrated snail darter. Anthropologists and ethnobotanists working in these areas indicate that many more useful substances are likely to be found in the primitive medical lore that has given the world quinine, reserpine and digitalis. A 1980 report of the National Academy of Sciences, entitled *Priorities for Research in Tropical Biology,* lays out a crash program to salvage as much scientific value as possible from the few years available in which to study these vanishing ecosystems.

Absence of adequate technology has been an important factor in the adoption of ecologically harmful development practices in many of these areas. To conserve these treasures requires research in such natural sciences as ecology and wildlife biology and the study of the preservation of materials. It also requires institutional innovations that further national and global aims by addressing local needs. For example, the careful planning of tourist facilities to enable local people to share in tourist revenues shows promise of permitting East African governments to conserve their great game parks both as assets to national development and as part of the world's ecological heritage.

In a historical and philosophical sense, perhaps, these treasures are the common heritage of all humanity. As a practical and political matter, this heritage will be lost unless its preservation is in the interest of the (usually hard-pressed) LDCs whose sovereignty prevails. If Americans wish these treasures to survive, it is in their interest to help developing countries to mobilize the technology needed for their preservation.

Rapid Population Growth and Unemployment. These issues are distinct in origin but together give rise to one of the most serious problems confronting the developing countries—one that also affects the United States both directly and indirectly. The urban unemployed and under-employed are among the chief sources of political instability in LDCs, many of which occupy geopolitical positions of strategic importance. The lack of employment opportunity in both rural and urban areas of these countries is the main pressure for illegal migration to the United States. Forced by rapid population growth and regressive systems of land tenure to extend their struggle for survival to marginally productive forests and deserts, the world's poor are exerting devastating environmental pressures in such areas.[7] These root problems cannot be bottled up indefinitely.

The population problem is ideally suited to the scientific and technological cooperation of both the public and private sectors and both the natural and social sciences. The development in the 1960s of such modern contraceptive methods as the pill, intrauterine device, and injectable contraceptive made the establishment of family-planning programs possible in LDCs. Such programs require a sensitive understanding of specific LDC needs; an understanding of the risks of using industrialized nations' approaches, standards and systems in inappropriate settings; and emphasis on the building of local capability. Major technological needs appear to be in biomedical research related to human reproduction and in efforts to develop and commercialize promising contraceptive technologies that have passed the research stage.

The unemployment problems of developing countries are based to some degree on inappropriate patterns of technological development. These are caused, in turn, by lack of local technological capacity, lack of information regarding technological alternatives, uncritical copying of the technologies observed in developed countries and government policies that overly protect against internal and external competition, encourage excessively capital-intensive investments and discriminate against small-scale industry.

In most developing countries, there is a crucial need for development strategies that create productive jobs. These require policies designed to encourage investment and, if necessary, develop technological alternatives that are both efficient and suited to labor-intensive operation. Such measures should be combined with policies to increase the demand for technology, to create and strengthen scientific and technological infrastructure and human resources, to build technological capacity in the productive sector and to remove incentives that bias technological development toward inappropriately capital-intensive solutions.

These are primarily matters of domestic policy, but there is ample room for international collaboration on the development of labor-intensive technologies. Research has shown that earth for civil works can be moved by large numbers of workers as efficiently as by machines, if proper attention is paid to the training of foremen, the nutrition of workers and the quality of hand tools. Similarly, agricultural engineers at the International Rice Research Institute in the Philippines have designed a variety of low-cost machines for the production of paddy rice and have assisted local firms to manufacture them.

Such research cannot by itself solve the LDC unemployment problem. But it can demonstrate that improved labor-intensive technology is feasible and can serve as the basis for industrial strategies that give equal weight to growth and the creation of productive jobs.

Geophysical Research. Research on the frontiers of oceanography and meteorology, which is critical to the development of long-range weather forecasts and to the location of undersea mineral resources, requires the full cooperation of developing countries. A major frontier of numerical climatology, the discipline that deals with the global circulation of the earth's atmosphere, lies in the understanding of the tropical atmosphere and its interaction with the oceans and with temperate regions. Research in this area is needed, both for long-range weather forecasts in temperate regions and for an understanding of the monsoons on which most tropical agriculture depends.

Many areas of critical interest to oceanographers lie well within the 200-mile limit of coastal jurisdiction of developing countries. These countries require assurance that the research vessels are not a cover for commercial or military espionage. More fundamentally, they should enjoy full participation in the gathering and interpretation of the data. Either goal would require a considerable strengthening of the technological capacity of coastal LDCs, a strengthening that is in the best interests of the oceanographic research community and, hence, of U.S. foreign policy.

Threats to the Global Environment. Several critical global environmental problems are beyond the ability of any one country to address individually and require a cooperative effort by all nations. Scientists now agree, for example, that burning fossil fuels and clearing forests will increase the carbon dioxide content of the atmosphere to the point where it will change the world's climate in a significant though unpredictable way over the next 50 years or so.[8] Moreover, through a complicated chemical sequence, the use of fertilizer may reduce the strength of the atmospheric layer of ozone that protects the earth from ultraviolet radiation.

Developing countries are major contributors to the loss of carbon in soil and standing biomass through deforestation, and several of the more advanced LDCs are, or soon will be, major users of fossil fuels and fertilizers. Both of these issues are rife with scientific uncertainty and require monitoring and research on a global scale.

The United States has direct interest in the participation of developing countries in global research efforts on these issues. In order to participate in such international efforts, LDCs must have the scientific and technological capacity to recognize that the world faces a serious problem, that their participation is important and that they need to participate in appropriate international research, monitoring and remedial measures. Although LDCs will be justifiably convinced that their primary need is for meteorologists and climatologists who can apply their skills to agriculture, transport and other more immediate problems, it should be possible to harmonize national and international areas of need.

Scientific and Technical Collaboration and Foreign Policy Goals. This broad survey of the scientific and technological aspects of U.S. foreign policy toward the developing countries has identified numerous areas where long-range cooperation could contribute to concretely defined foreign policy objectives and to important developmental goals. (Technological "stunts" designed for short-term impact on bilateral relations have been deliberately excluded.) Among the long-range goals:

- diversifying sources of oil for America and its allies;

- freeing developing countries from economic and political dependence on a small number of oil-exporting countries by encouraging energy conservation and the discovery and exploitation of renewable and nonrenewable energy resources;

- decreasing the long-run increase in world food prices by encouraging food production;

- alleviating problems of rapid population growth, unemployment and migration;

- conserving the world's cultural and ecological heritage;

- improving technology for exploiting undersea resources and for predicting long-term weather conditions;

- managing global environmental problems;

- improving health care technology and its availability.

Choices among these options depend on both objectives and preferred approach. How can one choose rationally between starving people, sick children, vanishing ecosystems and stable governments?

Of the global problems in which the U.S. stake is clearest, food and energy are of urgent importance and lend themselves to concerted international action. The accomplishments of the Consultative Group on International Agricultural Research (CGIAR) show the power of this approach in addressing the problems of both rich and poor. It is essential that U.S. support to CGIAR be expanded and that a global program for mobilizing energy technology be quickly developed and funded so that LDCs can take advantage of all appropriate sources of energy, both renewable and nonrenewable. (For further discussion of CGIAR, see Chapter 8.)

Another urgent international need is for action to study and conserve ecological and cultural treasures that will otherwise be lost forever. Programs in this area should begin immediately, but full mobilization will take some years to build up because of the scarcity of qualified personnel.

Population research is no less urgent and important than research on food and energy. A substantial international effort is already under way so that priorities for new international action are less clear. Given the certainty of limited political and financial resources in the early 1980s, major initiatives on population research may have to be deferred until some degree of consensus can be reached. The achievement of such consensus is thus an important policy objective. (Additional discussion of this topic will be found in Chapter 9.)

Science and Technology as a Dimension of
National Development

A succession of U.S. administrations has viewed the overall development of LDCs as in the long-run interests of the United States. In addition, the United States has, from time to time, taken a specific interest in the development of particular countries deemed of special geopolitical interest.

In either case, sound long-term development requires attention to a broad range of scientific and technological considerations. In particular, local technological capacity is a prerequisite to an effective attack at the national level on virtually any of the global problems noted in this discussion. For example, lack of indigenous technological capacity often leads LDCs to adopt sophisticated technologies used in developed countries, which frequently wastes scarce capital and foreign exchange, contributes to widespread unemployment and severe disruption of social traditions and leads to economic and political crises. Lack of technological capacity also hinders efforts to develop local energy sources. High rates of population growth, coupled with development policies that fail to encourage the use of employment-creating technology, fuel political unrest and the unwanted migration of workers.

The fields of science and technology are thus not the arcane preserve of specialists in the universities and research laboratories but are basic to development strategy. To the extent that U.S. foreign policy is concerned with the long-term health of specific developing countries or of the developing world in general, there must be a clear understanding of the scientific and technological dimension of development. This dimension includes not only research and development but the application of innovative technologies and the choice of technologies "off the shelf." It also encompasses the formation of human resources, the development of local capacities to adapt, absorb, create and use technology and the elaboration of national and sectoral policies designed to encourage technological innovation, assess its effectiveness and to guide it in socially useful directions.

The scientific and technological dimension of development further includes the capacity to adjust to advances in technology and to the technological consequences of such global trends as changes in the price of energy and other key commodities. Included as well is the capacity for informed participation in research and policy discussions on global technological and environmental issues.

As is clear from the discussion thus far, our definition of technology is deliberately broad—namely, the application of knowledge to achieve a practical objective. Thus defined, technology includes both equipment (hardware) and the institutions and management practices (software) needed for its effectiveness.

By this definition, technological development includes the evolution of the technology in use in a country as well as the country's development of its capacity to mobilize technology—i.e., to assess needs, resources and challenges and to choose, adapt, create and implement technology to meet defined objectives.[9] Such capacities can be found in universities, technological institutes, research laboratories, government agencies, private or publicly owned consulting firms and producer enterprises or small volunteer organizations. They require skills derived from both the natural and social sciences, and these skills are needed in some form at all stages of development.

In the early stages of technological development, a country must concern itself with the building of basic scientific and technological infrastructure—a university, minimal research facilities and a technical library, followed by scientific professional organizations and some form of national research council to administer fellowships, to coordinate research funding and to relate research to the country's needs. An agricultural research laboratory and the rudiments of industrial standards then follow.

Moravcsik and Ziman, in their classic article, "Paradisia and Dominatia," have described the difficulties facing the scientist from a developing country who, with his Ph.D. fresh from a leading U.S. or European university, returns home to discover that he must now assume responsibility for building a curriculum, a library, a workshop, a department and sometimes a national scientific community—all tasks for which he is completely unprepared—as well as carry on research with inadequate funding and with few of the support services that young U.S. scientists take for granted.[10]

The most difficult problem of technological development—that of relating the fledgling scientific and technological community to the mainstream of the economy—arises in its most elementary form at these early stages. Every developing country faces major decisions that cannot be deferred until it achieves a reasonable measure of technological capacity. Investment projects must be planned and development policies devised. The infrastructure inherited from the colonial past must be maintained and expanded to meet the demands of modernization. The past and probable future effects of the introduction of technology must be assessed, and the range of technology in use must be broadened.

Except for a few oil exporters, LDC resources, such as money, managers, trained people and institutional and physical infrastructure, are limited. Political leaders and economic planners are beset with pressing social, economic and political problems. Their staffs typically include able, highly trained professionals who are equal to the best in any country. But they are usually few in number and are thinly stretched over an enormous range of responsibilities. Moreover, they are rarely familiar with science and technology and neither provide substantial resources for research nor press for sorely needed technological development. There is also a large, unmet and urgent need for informed indigenous control of decisions concerning the overall conception and design of policies and projects, to insure that technological development choices are made with the full involvement, understanding and concurrence of local people and with local conditions and needs fully in mind.

The problems of undeveloped technological capacity are acute in developing countries that have come into sudden wealth through the export of petroleum. These countries must somehow invest vast sums quickly, under intense political, social and commercial pressure, while simultaneously developing local capacity, often virtually from scratch.

Thus defined, technological development lies at the heart of political stability and development strategy. To the extent that U.S. foreign policy is concerned with long-run development in LDCs, it should seek to foster a strong indigenous technological capacity where it is lacking. The development of local capacity should be an essential element of efforts to attack the more immediate and specific scientific and technological problems covered in the discussion that follows.

Many developing countries, on the other hand, have achieved a substantial degree of technological capacity. In these more advanced countries, the basic technological infrastructure is typically in place, but, in most, patterns of development have been based on the importation of foreign technology with little effort to learn to adapt it to the local situation, to reproduce it under similar conditions or to improve it. These patterns have left technological institutions isolated from the economy. In these countries, moreover, economic policies that affect interest rates, exchange rates and wage and tariff levels have frequently been responsible for a pattern of technological development that is inappropriate to local factor prices and that fails to create enough productive jobs.

Despite their relatively advanced level of technological capacity, these countries hold some of the poorest people in the world. Too often, social programs designed to improve their lives or develop the informal sector neglect to encourage scientific and technological research and innovation, resulting in little demand for the development of the simple, low-cost technology that could address their most pressing problems.

In the most advanced developing countries, scientific and technological research, development and innovation are important elements of market competitiveness. In some of their exports, these countries use up-to-date means of production, adapted to local conditions. They also use modern techniques to identify and serve their markets. Some have also begun to export technology, usually to other developing countries, in the form of capital goods, turnkey plants, licenses and technical services.

Conflict and Cooperation in Technological Development

U.S. foreign policy toward technological development in LDCs, especially those that are more advanced, is ambivalent. However much it may appreciate the role of technological capacity in long-term development strategy, the United States must consider a variety of issues in which its interests may conflict with those of specific LDCs. Such issues include commercial competition, the "export of jobs" through overseas investment or technology transfer, the threat or reality of nuclear proliferation, the supply and price of raw materials and the possible conflict between the desire of an LDC to develop its own industry and technology and that of the United States to export its manufactures, agricultural products and services.

These considerations are present to some extent in U.S. relations with technologically less advanced countries, particularly oil exporters, that are major powers in global resources politics, important markets for equipment and services and potential competitors in capital- and energy-intensive products like fertilizers and petrochemicals. Indeed, conflicts over such issues as controls on the import of toxic substances are most acute in countries that lack the technological capacity to draw up and operate an appropriate regulating mechanism.

As an additional complication many of these issues have entered the North-South dialogue in the form of demands for improved access to proprietary technology, international codes of conduct for technology transfer, buffer stocks to stabilize the price of commodities exported by LDCs and international arrangements for the exploitation of undersea minerals. These issues are pressed most vigorously in international diplomatic forums by governments of more advanced LDCs who seek to advance the interests of their growing modern industrial sector.

For the U.S. policymaker, these areas of conflict are awkward in three ways. First, they are sufficiently specialized that they rarely attract the sustained interest of high-level officials. Second, many concern areas of commercial interest in which the government is reluctant to become directly involved. Third, many of the remedies proposed by the LDC representatives may not, in fact, be the most technically effective ways to address the problem.

Although the U.S. response to these demands clearly involves more diplomacy than technology, the most constructive response would be to identify and study that portion of the problem where North and South have interests in common, to define its practical content carefully and to devise and promote a mutually beneficial technical solution. Whenever possible, this type of solution could be pursued, designed and implemented independently of the broader controversy. The purpose is not to win the diplomatic confrontation but to make progress toward solving the long-range problem. This approach seems more attractive than the alternatives, which are (with some oversimplification) either to "stonewall" or to offer a less expensive solution, which is demanded by the developing countries and which the United States has good reason to believe would not be effective. The following pages briefly outline several tentative applications of this approach to some of the thorny problems confronting U.S. diplomats in areas of science, technology and development.

Transfer of Commercial Technology. This issue has come to symbolize to the business community all the technological aspects of foreign policy toward the developing world. Developing countries have, through a variety of diplomatic forums, pressed for an easing of the costs and conditions of the transfer of commercial technology through a binding international code of conduct. Some have even asserted that knowledge should be a free good. Developed countries have replied that they do not have, nor wish to assume, control over such commercial transactions in the private sector, that commercially useful knowledge is costly to produce and deserves its full market value, that most of the technology needed by developing countries is available without restriction and even without charge and that any abuses can best be (and indeed are being) dealt with through national government regulations.

These disagreements are, in one sense, an extension of international business negotiations between suppliers and purchasers of technology. Many LDCs, in exercising their sovereign right to do so, have established regulations on the international commercial transfer of technology, which typically limit the size of royalties and prohibit certain provisions in commercial agreements. Agreements are often not allowed to ban the export of products made with imported technology or to require that raw material be purchased from the technology supplier. A regulatory body may be required to participate in the negotiations between suppliers and purchasers of technology.

At the pragmatic level, proponents of such regulations assert that they limit the foreign-exchange costs of technology transfers and eliminate onerous restrictions without hindering the flow of technology. Critics assert that supposed savings may simply be shifted to some other entry in the foreign-exchange outflow ledger or may be counterbalanced by the loss of benefits caused by project delays while the transfer agreement is being reviewed and approved. Only time will tell whether such regulation will in fact improve the terms of technology transfer without slowing it down.

In the meantime, there may well be unexplored avenues for collaboration in areas of mutual benefit. U.S. business has an interest in promoting the sale of technology, whether in the form of equipment, technical services, licenses or know-how. U.S. labor shares this interest, as long as these sales are matched by investments in domestic innovation intended to ensure continued U.S. competitiveness and with assistance to workers displaced from noncompetitive industries. Assuming that domestic innovation does not slacken, there is every reason to explore the possibility of creative mechanisms to encourage technological collaboration between U.S. and LDC firms.

Such measures already form part of U.S. bilateral agreements with Israel. A bilaterally funded foundation in that nation promotes technological collaboration between private firms in the two countries, typically but not exclusively providing for research and development leading to the commercial application of an Israeli technology using U.S marketing skills. Over the next several years, the same pattern could be extended to other relatively advanced countries, such as Korea, the Philippines, Taiwan, Thailand, Brazil, Mexico and Jordan. Comparable programs of cooperation might be arranged among trade or professional organizations. Another interesting suggestion is that of Jack Baranson, who has pointed out the need for a special facility to finance the front-end costs of a technology supplier in the United States who must incur expenses for technical services months or years before receiving royalties based on sales.[11]

Negotiations with Transnational Corporations. The wide areas of disagreement between developed and developing countries with respect to the activities of transnational corporations should not obscure their agreement on one fundamental issue: stable business agreements are in the interest of both sides, and the most stable agreements are those that are equitable. This premise leads directly to the somewhat paradoxical conclusion that it is in the direct interest of transnational corporations that their overseas counterparts be skilled in negotiations so that they may arrive at agreements that protect their interests and that they expect to fulfill. This interest extends to the U.S. government, not only because of its interest in U.S. commercial relations abroad, but because disputes with overseas investors constitute a major irritant in bilateral relationships with developing countries.[12]

Scientific and Technological Information. Developing countries have demanded improved access to scientific and technological information, which they find to be a near monopoly of the industrialized countries. This demand sometimes takes the extreme form of unfettered access to proprietary information. More recently, it has been embodied in a proposal for a cumbersome network of national focal points (too often serving as information depositories rather than information services) under United Nations auspices, which in this author's view will do little to aid the supposed beneficiary of the system, namely the user of technological information in the developing country.

The United States has an obvious interest in helping developing countries meet their technological information needs, especially when the technology is produced in the United States. To date, the major U.S. response to such demands has been a modest but useful program to improve access to the huge store of technical information in the National Technical Information Service (NTIS). Further efforts in this area would require the development of information networks at the national level designed to help the LDC users define and meet their own needs for information. Such systems could then develop means of access to the many existing data banks around the world. It would then be in the U.S. commercial interest to facilitate the access of such national services to any U.S. sources not covered by existing systems and to maintain a professional staff to handle specialized information.

Commodities. Developing countries are convinced that they are being victimized by low and fluctuating prices for natural commodities that they export. Although the United States has generally resisted strong diplomatic pressure for buffer stocks and other devices to alleviate this problem, it has supported research and development on many such commodities, which is much less expensive than buffer stocks and which is in the longer-term interest of both producers and consumers.

Careful attention to agricultural and technological research has allowed natural rubber to compete effectively with synthetic rubber, providing consumers with a useful engineering material and allowing Malaysia and other rubber producers the time and foreign exchange needed to diversify into other crops. This experience, and that of international organizations concerned with wool and cotton, has shown how research, integrated with marketing and promotion in a commercially oriented strategy, can defend the market competitiveness of a natural commodity against intense competition from synthetic substitutes. By contrast, neglect of modern marketing techniques and inadequate agricultural and industrial research have been primary factors in the rapid decline of the market for jute, a principal export of Bangladesh and India.

The United States is a charter member and major supporter of the International Institute for Cotton (IIC), an intergovernmental organization for the defense of the market competitiveness of cotton through industrial research, promotion and service to the textile trade. The United States has also supported the proposal for a Cotton Development International (CDI), which would absorb the institute's program and extend it into agricultural research and into a more active role in increasing the amount of cotton used by the LDC textile industry.

The CDI proposal, made public in 1976, is still under discussion by governments. Unfortunately, it has been considered a subsidiary issue in discussions under way at the United Nations Council on Trade and Development (UNCTAD) for cotton buffer stocks—which many negotiators feel is the "real" issue, regardless of the fact that the technological defense of market competitiveness can achieve major results at much lower cost. Consequent delays have threatened the future, not only of CDI, but of IIC itself.

The history of the CDI proposal illustrates the difficulties faced by constructive attempts to develop practical, mutually beneficial programs of scientific and technological cooperation in this highly politicized area. Nevertheless, such efforts deserve U.S. support, especially when they concern commodities, such as cotton and jute, that contribute heavily to the livelihoods of poor people in LDCs.

Communications. LDCs exert major diplomatic leverage in two international organizations that have a substantial influence over global communications policies of importance to the United States. First, the United Nations Educational, Scientific and Cultural Organization (UNESCO) is engaged in a major debate over measures to rectify the imbalance in information flows between developed and developing countries, which are convinced that journalistic coverage by private news services is, in some cases, insensitive and displays insufficient understanding of their problems.

The United States fears that the measures proposed by LDCs to UNESCO could legitimize the efforts of governments to control the flow of news from their countries. Additionally, the International Telecommunication Union is responsible for convening intergovernmental conferences to allocate radio frequencies among conflicting uses—a function of both military and civilian significance. U.S. diplomacy in both cases suffers from the absence of any program to assist developing countries with their communications problems. Yet, this is an area where the most advanced U.S. technology has clear application to important development problems that would be difficult to address in any other way.

Substances Involving Hazard. U.S. environmental groups have occasionally proposed that the export of such substances be subject to the same restrictions as their domestic use. The balance between risks and benefits in developing countries differs greatly from that in the United States, however. The United States would assume an impossible burden if its courts had to decide, for example, whether DDT should be used to control malaria in Sri Lanka or whether insecticides are being properly applied to cotton crops around the world.

On the other hand, many developing-country governments cannot deal effectively with the pressures, both internal and external, to allow misuse of these useful but dangerous substances. Toxic substance exporters often help LDCs draft codes of control but may do so in a self-serving manner. There have been instances of efforts to influence foreign governments to use pesticides improperly or excessively and of warning labels that are unreadable by foreign users or lack essential cautions. Although there is no perfect solution to this problem, it would be to everyone's advantage to help LDCs build up their own capacity to deal with these matters and protect their own interests.

An analogous situation exists in the field of pharmaceuticals, where there is an obligation—not always perfectly honored in practice—to warn physicans and consumers of possible side effects and contraindications. Here, too, risk-benefit factors may be different in an LDC than in the United States. A country in the midst of an epidemic may be willing to license a vaccine or drug with significant side effects, even though there would be no need to accept these risks in the United States where the disease is virtually nonexistent.

This issue has become particularly acute in the case of the injectable contraceptive Depo Provera. Depo Provera has not been approved for contraceptive use in the United States because of studies that suggest a possible link with cancer and other side effects. Numerous LDCs nevertheless use this substance and request that the United States provide it, arguing that their population problems are urgent and that the risks of pregnancy and childbearing in their countries far outweigh the drug's possible hazards.

It is a mistake to impose U.S. conditions on LDCs and to argue, as some U.S. groups have done, that drugs such as Depo Provera should not be exported because they carry risks unacceptable in this country. In the end, it is the LDCs themselves, through development of their own regulatory capacities, who will decide the appropriateness of any imported technology.

Current U.S. Policy and Programs

I have not attempted a comprehensive review of U.S. programs for scientific and technological cooperation with developing countries or of the policies that underlie them. What follows is a brief overview of existing policy and practice, with particular emphasis on bilateral programs, for the purpose of comparison with the approaches recommended in this paper.

For convenience, I distinguish four broad, and in some cases overlapping, mechanisms for U.S. bilateral scientific and technological cooperation with LDCs:

1. Bilateral agreements with countries of geopolitical importance;
2. Development assistance programs;
3. Extension of domestic programs into the international sphere;
4. Programs set up in pursuit of global policy objectives.

Bilateral Agreements

The United States has signed agreements of bilateral cooperation in science and technology with (then) developing countries as diverse as New Zealand, China, Egypt, Saudi Arabia and Spain. Many of these agreements have resulted in large and important programs.

From the point of view of U.S. foreign policy toward the developing countries, the chief function of such agreements has historically been to improve the atmosphere of bilateral relationships. Such efforts have followed a prescribed order: first, the exchange of athletic and cultural attractions; then, the scientific mission to arrange a cooperative agreement; then, the addressing of "real" issues. The objective of the scientific mission was to create some cooperative effort between the United States and the other country; the subject of cooperation was unimportant and could be left to the scientists, who typically chose subjects where research capacity in the cooperating country was strong, regardless of its relevance to national needs. Conversely, if bilateral relationships were chilly and a thaw was considered undesirable, there has been little or no provision for scientific and technological cooperation.

Recent bilateral agreements with China and several African countries show an encouraging shift from this pattern, at least in the choice of the subject for cooperation. Formal agreements were preceded by surveys of each country's needs to identify the best subjects for bilateral cooperation. These agreements ranged well beyond research to include, for example, collaboration in water resources planning in major Chinese river basins.

Another interesting bilateral experiment is the U.S.-Israel Binational Foundation for Industrial Research and Development, mentioned earlier, which encourages commercially motivated, enterprise-to-enterprise technological cooperation in the private sector. The work of this foundation provides, to this author's knowledge, the only example in the U.S. federal government of direct support to industrial research and development awarded on purely commercial criteria.

Development Assistance

Science and technology have formed part of U.S. bilateral development assistance for many years. Assistance programs have supported, for example, research on human reproduction, forestry, water resources, pest control and tropical diseases. These assistance funds have also supported a useful series of publications by the National Academy of Sciences that convey, in compact form, the state of knowledge in relatively unexplored fields of science and technology that promise applications of high economic potential in LDCs. Examples include ferroconcrete, unexploited species of tropical legumes and fast-growing trees.

Since 1973, the objective of the foreign aid program has been to help meet the needs of the desperately poor for food, health, education and, in recent years, fuelwood. This humanitarian objective has been a prerequisite to continued congressional support for the development assistance budget and has inspired efforts critically important in the global struggle to alleviate poverty.

Assistance programs under this new policy support the development of research capacity in these fields in the poorest LDCs, which are now the sole recipients of development assistance. In addition, a major new group of programs, called the Collaborative Research Support Programs, supports the building of U.S. capacity to collaborate with food and nutrition researchers in LDCs.

The shift in emphasis to the problems of the poor has also increased interest in low-cost technology. Since much of the technology used to solve comparable problems in industrialized countries is far too expensive for poor LDCs, development assistance agencies have been forced to consider innovative solutions to such problems as health services, nutrition and basic education, in order to meet these needs at a cost low enough that the approach can be extended to large numbers of people within the resources available to the developing country.

Several U.S. congressmen have indicated a further special interest in "capital saving" technology as the key to meeting the needs of the poor and have added provisions to foreign aid legislation to ensure the use of such technology. This has prompted support to community action groups in LDCs capable of applying such technology to the needs of the poor through the Agency for International Development (AID), the Inter-American Foundation and Appropriate Technology International.

These efforts have been useful correctives to the tendency of development assistance agencies to apply familiar technologies, even when these are unsuited to the problem at hand, and have given new legitimacy to technologies that might otherwise have been regarded as unworthy of a modern country. On the other hand, they run the danger that "appropriate technology" may be rejected as second-rate or dismissed as the latest fad or panacea without proper consideration of its merits.

The United States has made major contributions to multilateral agricultural research on food crop production and to biomedical research in tropical diseases and human reproduction. It has also given substantial support to building the needed capacity for carrying out research on these problems. The food crop research financed by CGIAR has been an outstanding success. As a funding mechanism, the group is already serving as a model for the financing of research in fields that lend themselves to integrated global programs under international management and control. Such research has sometimes taken place in international centers but, in other cases, has taken the form of a network of research institutions in developed and developing countries, such as that of the Integrated Program of Training and Research on Tropical Diseases of the World Health Organization.

International Extension of National Programs

Many programs developed for scientific and technological research in the United States have been extended to other countries. The National Aeronautics and Space Administration (NASA) has supported experimental applications of satellite techniques of remote sensing in developing countries; the Communicable Disease Center and the Department of Agriculture have, respectively, investigated diseases and insects that posed threats to the United States; and the Geological Survey has studied earthquakes and volcanic eruptions in foreign countries to provide insight into similar events in the United States.

All of these programs provide technical assistance to their LDC counterparts, although more as a by-product than as a primary objective. Increased attention to this technical assistance could, at relatively low cost, greatly increase the effectiveness of these programs in building LDC problem-solving capacity and contribute to the global stock of knowledge on these subjects of worldwide interest.

Programs in Support of Global Objectives

By far, the major thrust of U.S. pursuit of global policy objectives has been in programs to abolish poverty. But there are also several U.S.-supported programs that deal with issues of global significance not directly connected with poverty.

U.S. participation in the Global Atmospheric Research Program of the World Meteorological Organization is primarily intended to provide the scientific basis for long-range predictions of U.S. weather. This objective can only be fulfilled through international cooperation, as it requires a global effort to fill major gaps in meteorological (and, to a lesser extent, oceanographic) data. Many of the most important of these are in the tropics. For this reason, U.S. participation in this program, although not primarily intended to assist developing countries, provides tens of millions of dollars for the study of the tropical atmosphere, a subject of critical interest to these countries.

Examples of efforts intended to fulfill global objectives are the international programs of the Department of Energy. The department has published long-range assessments of the energy needs of Peru, Egypt and several other countries. While the prime objective of this program is to slow the spread of nuclear power, its executors quickly realized that they had no choice but to try to take the point of view of their "customers" and to provide them with full assessments of all their energy options, including nuclear power. Such an approach, it is hoped, will discourage nuclear projects that are motivated by prestige and that lack techno-economic justification. On the other hand, these studies have not attempted to build local capacity for needs assessment and have sometimes tended to project U.S. conditions and requirements onto the developing countries without fully evaluating the alternatives.

General Assessment

It has long been apparent that U.S. programs of scientific and technological collaboration suffer from fragmentation, omissions and lack of funding or institutional support within the government. As Eugene Skolnikoff has pointed out in his "synthesis" paper in this volume, it has proven very difficult to fund programs that can neither compete at par with the best of American science, nor be justified as aid to rural development in the poorest LDCs. These are important priorities but surely do not span the whole set of objectives of U.S. foreign policy or of the overseas aspects of U.S. technology policy.

As early as 1971, the National Academy of Sciences recommended the establishment of an International Development Institute, separate from AID, as a focus for scientific and technological cooperation with LDCs as distinct from the transfer of resources. This proposal was revised and brought up to date after a thorough study by the Brookings Institution of the bureaucratic difficulties faced by AID in its attempts to support scientific and technological programs.[13]

In 1978, building on the Brookings study, the White House Office of Science and Technology Policy proposed an Institute for Scientific and Technological Cooperation (ISTC) as a mechanism to encourage and coordinate scientific and technological cooperation with LDCs. Establishment of ISTC was to fill several gaps in the institutional framework for dealing with technological problems in LDCs: to increase support to research and development, free of the pressure of the immediate priorities of AID programs and of the state of bilateral relations with cooperating countries; to make possible research support for projects not directly related to "basic human needs"; and to develop programs of scientific and technological cooperation with middle-income countries such as Brazil, Mexico and Korea, which are "graduates" of aid programs and on the way to becoming major industrial powers.

The ISTC proposal reflected increased understanding of the complex role of science and technology in development—as an important dimension of development rather than a "fix," a panacea or a stunt. It recognized that technology covered a broad range of levels of sophistication, from satellite-based remote sensing to improved clay and sand cookstoves. It recommended that sociological and institutional constraints to the diffusion of improved technology be explicitly addressed and that particularly complex problems—such as nutrition and public health—be addressed by integrated research on technology policy and institutional design. Finally, it gave developing countries a role in the management of the institute through the establishment of an advisory council on which they were to be represented. All of these were important advances.

Congress has not approved the establishment and funding of such an independent ISTC, allowing instead only a small increase in the budget for science and technology within the regular AID structure. AID has established the position of scientific adviser in the Office of the Administrator and is making arrangements to fund research on topics identified by the National Academy of Sciences through its Board on International Science and Technology for Development.

This is not the place to assess the prospects of some version of ISTC for eventual enactment or to review the efforts of its planners, who worked under difficult pressures of time and politics to design it and to justify it to the Congress. Some of the participants in that effort have suggested in personal conversations with the author that the difficulties were no more fundamental than the failure to convince skeptical members of Congress that the laudable objectives of ISTC, to which the Congress was basically receptive, could not be achieved within existing organizations. A few are even convinced that scientific and technological cooperation will become the major thrust of U.S. bilateral development assistance over the next several years, partly because science and technology are the areas in which the United States has the most to offer and partly because this kind of cooperation is cheaper than financing large investment projects.

Others, by contrast, cite the uneasiness of politicians at the premise that science and technology can fully contribute to LDC needs only if scientists are given a greater voice in such efforts and are able to use resources without the constraints of existing organizations.

In any case, the refusal of the Congress to appropriate funds for the ISTC or to contribute to the U.N. Interim Fund,[14] combined with the increasing unpopularity of foreign assistance—as shown by the annual difficulties faced by the foreign aid bill in Congress—show clearly the present lack of a strong domestic political constituency for improved scientific and technological cooperation with the developing world. The author would suggest that it is time to seek to create or strengthen the domestic political constituencies for such activities.

A likely basic constituency would seem to be the U.S. scientific and technological community. However, with the exception of a relatively small number of professionals concerned with scientific and technological development in LDCs, there is an absence of strong interest in this field among much of the leadership of the U.S. scientific and technological community. Proposals for support of research on problems of interest to LDCs, whatever their intrinsic scientific interest or practical importance, have been regarded as competition for budget resources with the "real" interests of the U.S. scientific community—namely, those problems defined by purely domestic interests of the United States.

There has been some change in this attitude, partly because of the expansion of scientific and technological cooperation with China, the Middle East and (to a lesser extent) Africa. Technical journals such as *Science,* the *New Scientist* and the *Bulletin of the Atomic Scientists* have begun to devote increased space to issues concerning developing countries. But there has been far too little effort to convey both the human importance and the intellectual challenge and excitement of the scientific and technological problems confronting the developing world.

U.S. scientists should, in the name of devotion to the pursuit of knowledge, assume some responsibility for the professional survival of their colleagues in LDCs. They should encourage U.S. funding agencies to make available the relatively small sums needed to keep scientific research alive in these countries, and they should acquire the knowledge needed to be able to prepare students from these countries for the special problems they will face on their return home.[15]

Popular interest among Americans in the scientific and technological problems of LDCs has been limited to subjects that echo the domestic concerns of organized groups. Public-interest groups have rendered a useful service in spotlighting the particular technological problems faced by women in developing countries, as well as the need to protect endangered species from commercial exploitation.

Occasionally, however, the projection of domestic political issues onto the different conditions of LDCs results in a somewhat distorted perspective. I have already discussed the difficulties of applying arguments designed for U.S. conditions to the problems of pesticide and contraceptive use in LDCs. There are many similar policy issues. The use of infant formula is heavily promoted to the poor in LDCs who cannot afford and frequently misuse it. Yet, such products are as essential to working mothers there as they are here. Here public pressure to discourage irresponsible advertising can have a useful effect.

Wages, working conditions and safety and environmental safeguards in most developing countries are far below U.S. standards. In many cases, scandalous conditions could be improved at little cost. Yet, LDCs cannot afford the unquestioning application of the standards typical of advanced countries. Here the efforts of U.S.-based labor unions to raise the awareness of their colleagues in LDCs are far more effective than calls for protection against "cheap labor."

By comparison, U.S. environmental and consumer groups have been curiously inward-looking. Americans mobilized to save the snail darter, yet they have paid little attention to the predictions that hundreds of thousands of species may become extinct in coming decades without having been catalogued, let alone studied for possible economic value or scientific interest.[16] Popular support is needed for research programs to survey and study existing flora and fauna in these areas and for ecologically sustainable strategies for their protection; in several poor countries these areas occupy much of the remaining unused arable land.

Public pressure might also be useful in persuading timber companies to adopt sustainable approaches to forest exploitation, even in countries where this approach may not be scrupulously required or even encouraged by local authorities. Such an approach might well be made unofficial U.S. government policy and urged on U.S.-based companies — much as foreign policy officials occasionally urge financial support of a shaky government of special geopolitical importance.

The distinguished Dutch economist Jan Tinbergen has pointed out the natural alliance between consumer advocates in developed countries, who seek to lower costs and improve the quality of products in the marketplace, and the advocates of increased trade with LDCs, whose products tend to be at the low-cost, low-quality end of the spectrum. Their combined efforts could lower prices of an entire array of products. Yet U.S. consumer groups have thus far spent surprisingly little of their political capital in opposing protectionism. It is strongly in the interest of the U.S. consumer both to insist on adequate measures to ensure innovativeness in U.S. industry *and* to place no obstacles before—and, in some cases, to assist—the technological capacity of LDC industry.

Conclusions

A major expansion and redefinition of scientific and technological cooperation with the LDCs is needed, not only for humanitarian or charitable reasons but also to address major concerns of U.S. foreign policy for the 1980s: food, energy, global political stability and the future of the world environment. The need for greater cooperation is especially acute in food, energy and population. Effective bilateral and multilateral programs are already under way in food crop production research. These deserve continued support and expansion. In addition, there is an urgent need for international efforts to assist developing countries to develop indigenous sources of energy. Such a program could be readily designed and implemented.

The urgency and importance of the population problem is no less acute, but the priorities for international technological collaboration in this field are less clear and require further study. This should be carried out without delay so that effective action can be undertaken.

Needs have also been identified for international scientific and technological collaboration in the study of parasitic and diarrheal diseases and for the study of ecosystems of great economic potential, such as the humid tropical rain forest, which are in acute danger of disappearance. Finally, there is a general need to support local capability to mobilize science and technology at the national level as a part of overall national development.

These suggestions, taken individually, may be relatively modest, but they add up to a substantial redefinition of U.S. attitudes and interests vis-à-vis the role of science and technology in foreign policy toward the developing world.

Although scientific and technological cooperation along the lines suggested in this paper do not necessarily give rise to technological spectaculars, they do directly affect U.S. voter interests in lower food costs, freedom from petroleum cutoffs, secure supplies of minerals, the continued expansion of the world economy and the expansion of world demand for technologically sophisticated U.S. equipment and services. They are essential parts of any strategy to eradicate the worst aspects of poverty and to conserve the global environment. They are important to any strategy to assist the long-run development of LDCs as a whole, or of such specific countries as the United States wishes to support for strategic or other reasons. And they are intrinsically challenging and exciting at a time when U.S. popular interest is returning to scientific and technological advances.

These facts can be used as the basis for efforts to expand public support for scientific and technological collaboration with LDCs, as well as to gain support within the U.S. scientific and technological community for such work. Policies and programs based on shared long-term goals should complement, not replace, current policies based on humanitarian concern for the poor in developing countries.

There is a need to continue, and in some cases to expand, research on small-farmer agriculture, forestry, renewable energy, parasitic diseases, low-cost housing and sanitation and other fields of specific interest to the poor. However, political support on this basis is palpably diminishing. In any case, purely humanitarian concerns do not provide a satisfactory basis for dealings with oil-exporting or middle-income developing countries, or indeed with the middle class of the poor developing countries who, after all, are the holders of political power.

There is no reason for LDCs to fear a U.S. policy based on self-interest. On the contrary, this is their best reason to hope for a consistent policy. It is unreasonable and unrealistic to expect any country to pursue for long a policy that does not derive from its own interests. The incoming administration has emphasized its intention to put U.S. interests at the center of its foreign policy.

A fully coherent and integrated policy toward the technological development of the developing countries no doubt must await the clarification of public attitudes toward the technological development of our own country. It would clearly be useful to build a consensus in this area, as the most important international technological issues vis-à-vis LDCs are but one aspect of a broader debate on the response of the United States to its interdependence with the rest of the world. This response requires a substantial effort to refurbish U.S. competitiveness and innovative capacity, which in turn will require a major rethinking among labor, management, consumers and the public. An interesting step in this direction was recently taken by the Economic Policy Unit of the United Nations Association of the United States, in its report entitled *The Growth of the U.S. and World Economies Through Technological Innovation and Transfer.*[17]

The world of the 1980s is small, interdependent and fragile. U.S. security depends on economic stability and growth, here and abroad; a diversified supply of resources from many parts of the world; and preservation of the global environment. It also depends on relief of the misery of poor people, development of productive employment opportunities and control of population growth. It may well be necessary to redefine public and official concepts of the national interest, which up to now have tended to refer primarily to military and strategic concerns.

All of these goals require that LDCs build the technological capacity to become full members of the international community. Their technological development deserves an important position on the foreign policy agenda of the United States. U.S. resources are plentiful and can readily be mobilized. Although U.S. initiatives have achieved major impact, they continue to fall far short of the efforts that should be undertaken in our own interest.

Notes

1. Council on Environmental Quality and the Department of State, *The Global 2000 Report to the President: Entering the Twenty-first Century,* 2 vols. (Washington, D.C.: Government Printing Office, 1980), vol. 1, pp. 1-2.
2. John K. Fairbank, *The United States and China,* 4th ed. (Cambridge: Harvard University Press, 1979).
3. World Bank, *World Development Report* (Washington, D.C., 1978), p. 33.
4. For a global overview of the common interests of the United States and the developing countries in scientific and technological collaboration, see J. Ramesh and C. Weiss, eds., *Mobilizing Technology for World Development* (New York: Praeger Publishers, 1979).
5. John M. Kalbermatten, "Appropriate Technology for Water Supply and Sanitation: Build for Today, Plan for Tomorrow," lecture to the Institute of Public Health Engineers, 18 November 1980.
6. Gautam S. Dutt, "Field Evaluation of Woodstoves," paper prepared for Volunteers in Technical Assistance, Washington, D.C., 19 February 1981, p. 5.
7. Overcrowding by poor people in need of housing also poses a major danger to urban historical treasures, such as the great traditional Arab market of Fez, Morocco.
8. Council on Environmental Quality and the Department of State, *The Global 2000 Report to the President,* vol. 1, pp. 36-37.
9. A fuller definition of technology is given by the Economic Policy Council of the United Nations Association of the United States:

Technology means human knowledge used to achieve ends. Thus it includes everything from the manufacture of computers to the marketing of breakfast cereals, and from the use of public health measures to prevent disease to the growing of more abundant and better crops. It also includes the knowledge of how to organize society in order to achieve desired ends (corporations, educational institutions, health care organizations) and how to cope with related consequences.

Economic Policy Council of the United Nations Association of the United States of America, *The Growth of the U.S. and World Economies Through Technological Innovation and Transfer* (New York: United Nations Association of the United States of America, Inc., 1980), p. 13. The author has defined technological capacity as:

The ability to make technological decisions that will influence the allocation of resources and the efficiency of production units. . . . [This goes well beyond technological infrastructure, by which term] we understand the public and private institutions which provide technical services for the selection, adaptation or creation of technologies, for pre-investment and project implementation work, for quality control and trouble-shooting in production processes and for technological management and planning both at the macro and the micro level.

Charles Weiss, Mario Kamenetzky and Jairam Ramesh, "Technological Capacity as an Element of Development Strategy," *Interciencia*, vol. 5 (March-April 1980), p. 97.

10. Michael J. Moravcsik and J. M. Ziman, "Paradisia and Dominatia: Science and the Developing World," *Foreign Affairs*, vol. 53 (July 1975), pp. 699–724.

11. See Jack Baranson, *North-South Technology Transfer: Financing and Institution Building* (Mt. Airy, Md.: Lomond Publications, 1981), pp. 143–146.

12. C. V. Vaitsos, "Government Policies for Bargaining with Transnational Enterprises in the Acquisition of Technology," in J. Ramesh and C. Weiss, eds., *Mobilizing Technology for World Development*, pp. 98–106; J. N. Behrman, "International Technology Flows for Development: Suggestions for U.S. Government and Corporate Initiatives," in Ramesh and Weiss, pp. 118–127.

13. Lester Gordon, Courtney Nelson et al., *Interim Report and Assessment of Development Assistance Strategies* (Development Discussion Paper #74, July 1979; obtainable from Harvard Institute of International Development, Harvard University, Cambridge, Mass.).

14. The proposed ISTC formed the centerpiece of the U.S. position at the U.N. Conference on Science and Technology for Development held in Vienna in August 1979. U.S. participation in the conference was constructive and critical to even the modest degree of success achieved by that conference. The main tangible result of the conference was the creation of the U.N. Interim Fund on Science and Technology for Development, a multilateral fund to support the development of scientific and technological capacity in developing countries.

15. For numerous ideas on this subject, see Michael J. Moravcsik, *Science Development: Toward the Building of Science in Less Developed Countries* (Bloomington, Ind.: PASITAM, n.d.); and Charles Weiss, "Science for Development: What Individual Scientists Can Do," in V. Rabinowitch and E. Rabinowitch, eds., *Views on Science, Technology and Development* (New York: Pergamon Press, 1975), pp. 137–143.

16. Here a special exception must be made for the World Wildlife Fund, which is sponsoring a unique experiment in ecology to determine the minimum size of ecological reserve suited to the conditions of the Brazilian Amazon.

17. Economic Policy Council of the United Nations Association of the United States of America, *The Growth of the U.S. and World Economies*.

8
U.S. Agriculture in the Context of the World Food Situation

Sylvan H. Wittwer

Introduction and Background

Agriculture is the world's oldest and largest industry and its first and most basic enterprise.[1] More than half the people in the world live on farms. Food is first among our needs. Events of the past decade have focused attention on a new element of national power and safety—the control of vital resources, one of which is food. Renewable agricultural production will become increasingly important in resource bargaining. The potential of agricultural production as a strategic resource internationally and within the domestic economy is under review.[2]

The renewability of the products of agriculture comes as a result of "farming the sun." Agriculture, through the production of green plants, is the only major industry that "processes" solar energy. Green plants are biological sun traps. The aim of agriculture in crop production is to adjust plant species to locations, planting designs, cropping systems and cultural practices to maximize the biological harvest of sunlight by green plants to produce useful products for people. Many products of agriculture may be alternatively routed as food, feed, fiber or energy. Plants contribute to world food production 94 percent of the total edible dry matter by weight. Animals contribute 6 percent. Most animal products are derived in turn from plants.

Achieving an adequate and secure food supply for all people is both a humanistic goal and a mark of progress. This paper focuses on science and technology as they relate to these goals, which are by no means easily managed or predictable. In the mid-1960s, for example, a two-year drought in India and Pakistan brought catastrophic shortages of food. The trend

Sylvan H. Wittwer is assistant dean, College of Agriculture and Natural Resources, Michigan State University, East Lansing, Mich.

was reversed by a green revolution in the late 1960s. Poor harvests in 1972–1974, however, produced a new surge of despair. This was followed by a wave of optimism in the late 1970s, brought by surpluses, low prices and record production. Finally, in the early 1980s, we again face prospects of worldwide shortages and runaway food prices.

More than 70 percent of the current world population (4.3 billion) and 85 percent of the projected population growth by the year 2000 are found in the less developed countries. A large part (80 percent) of the absolute, as well as the relative, poverty is found in the rural and agricultural sectors. Many of the rural poor in developing countries are landless laborers or small farmers with insufficient land and capital to earn an adequate living from farming.

Several hundreds of millions of people are chronically malnourished. More than half are children, and more are women than men.[3] Food production must be increased considerably in the future, or food and nutritional problems will become worse. According to the report of the Steering Committee for the World Food and Nutrition Study,[4] it will be necessary to increase food production by at least 3 to 4 percent per year between now and the beginning of the twenty-first century for significant improvement to occur.

These predictions are sobering in view of the trend during the 1970s for yields of the major food crops to reach a plateau both in the United States and in the rest of the world. Some of the possible causes for that leveling out have been outlined in the literature on the biology of crop production.[5] Meanwhile, energy-intensive farm inputs have risen sharply and continue an upward trend. These are ominous signs because the timetable for doubling of food production to meet estimated consumption in most developing countries allows only 7 to 10 years.[6] The decades of the 1950s and the 1960s were truly the "golden age" for gains in U.S. agricultural productivity (Figure 8.1). The yield fluctuations during the 1970s suggest that the consistent gains in the two previous decades are not likely to be repeated.

Increases in food demand will come from both growing populations and increases in consumer incomes. The major force in the growing commercial demand for food is rising affluence. Expanded productivity per unit land area, per unit time and per unit cost is the primary source for the projected 3 to 4 percent yearly production increases needed. According to the Food and Agriculture Organization of the United Nations, 28 percent of these increases could come from an expansion of arable land — with a progressively decreasing portion in time — and 72 percent from intensification of land use through higher yields and increasing the number of crops produced per year.[7] In contrast, the *Global 2000* Report projects that world food production will increase 90 percent over the 30 years from 1970 to 2000.[8] It also

Figure 8.1. Composite Index of Crop Yields for the State of Michigan (1880-1980) and for the United States (1910-1980). (1967 = 100)

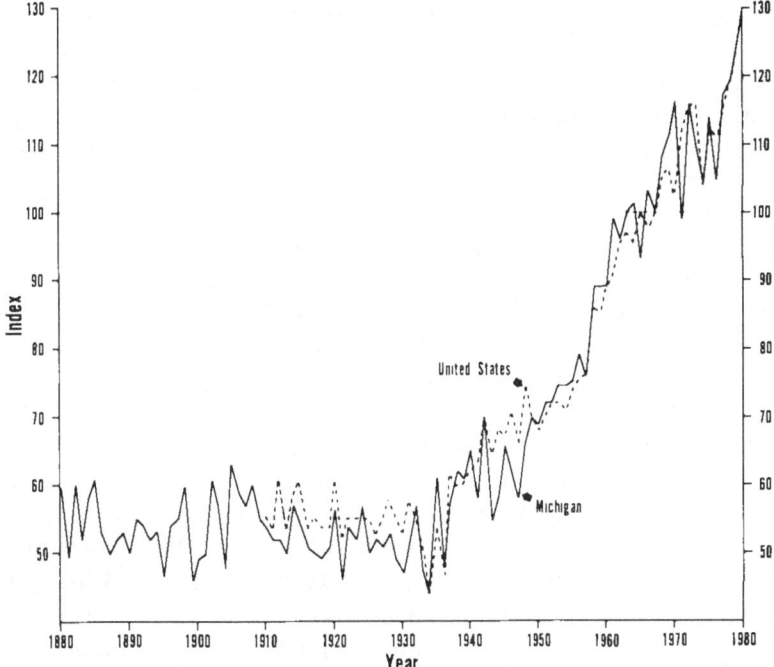

Note: There was no consistent gain from 1880 to 1940, there was a precipitous rise during the 1950s and 1960s, and there was a tendency to plateau in the 1970s.

Source: Author, with the assistance of Karl T. Wright, Professor Emeritus of Agricultural Economics at Michigan State University.

projects that arable land will increase only 4 percent by 2000 and that most of the increased food output will have to come from higher yields. The key to sufficiently large and sustaining yield increases will be technological change.

The following scenario is likely. Larger populations, greater affluence and increasingly greater consumption of animal proteins will intensify pressures for more intensive cultivation of available land. The pressure on land will be accentuated by the relative scarcity of water as its use approaches the limit of potential supply. The scarcity and expense of energy will then further aggravate the situation.

The challenge will be to "make two ears of corn or two blades of grass to grow upon a spot of ground where one grew before" (Jonathan Swift in *Gulliver's Travels*). This can be done by increasing traditional inputs but at greater costs. The challenge will be to increase inputs at less cost, so that food prices can be maintained at reasonable levels. To achieve this policy strategy, one must take into account resource inputs, both natural (climate, land, water) and manmade (energy, fertilizer, pesticides, human labor, machinery). Their costs, availability and renewability must also be taken into account.

In the United States the development of labor-saving technology has been a significant goal and achievement. Never have so few people produced so much. A farm worker's production can be measured by the number of people, in addition to himself, he can feed. In 1980, one farm worker can feed 60 other people; in 1970 he could feed 30 other people; in 1940, only 10 others; and in 1900, only 6 other people. The increased productivity per unit of labor input may be attributed to more extensive and skillful use of the resources of water, energy, fertilizer and pesticides. It is also the result of better management, more timely operations and more efficient and productive equipment. Mechanization in the United States has enabled farmers both to carry out their field work on a timely basis and, at the same time, to allow for management activities. Mechanization in the United States has been a necessity because of the unavailability, uncertainties and rising costs of human labor. In Japan and some other industrialized nations, where land, water and energy resources have been limited and labor more plentiful, mechanization has not been so prevalent. In these countries, yields are higher, but output per farm worker is much less.

Thus, there are two general types of food production technologies for the future:

- food production, based on a high degree of mechanization, with extensive use of land, water and energy resources, and little use of biologically based technology;

- food production, based on biological technology and sparing of land, water and energy resources.

The future will show a national and worldwide shift from a resource-based agriculture to one based on biological and scientific technology. The emphasis will be on raising output for each unit of resource input and on easing the constraints imposed by relatively inelastic supplies of land, water, fertilizer, pesticides and energy.

Reasons for the Shift to Biologically Based Agriculture

Ruttan has pointed out that the shift to a biologically based agriculture has already occurred during the first part of the twentieth century in Japan and certain European countries.[9] Whereas the United States has followed a mechanical-resource intensive technology, Japan has followed a biologically and chemically based technology which is sparing of resources. Incentives for increasing yield technologies have lagged until recently in the United States, compared with Japan and some European countries, because of the abundance and low cost of resources in the United States. But this situation is bound to change in the United States and elsewhere. It is projected that almost all future increases in food production will be a result of increases in yield (output per unit land area per unit time) and from growing additional crops during a given year on the same land. There are really no other viable options.

Any new agricultural technologies for the future will combine more dependable production and higher yields and will emphasize strategies that are more labor intensive than capital intensive and that spare rather than exploit resources. They must be nonpolluting and scale neutral (adaptable to any size of farm). They must increase the demand for underutilized labor resources. And they must offer solutions to the following global problems:

- Poverty
- Inflation
- Malnutrition
- Underemployment
- Deforestation
- Soil erosion
- Changing climate
- Communication gap between agriculturists and policymakers
- Uncertain responses of political institutions
- Population increase
- Shortage of firewood
- Water-logging and salinization
- Uncertainties of energy supplies
- Toxic chemicals in the environment
- Unstable production and yield
- Grain-food/energy conflicts

Technologies of this sort already exist. Some will be described later.

One of the constraining myths in setting the food research and technology agenda for the future is the belief that all we have to do is put to use the technology we now have and all will be well. This implies that we do not need more research but only better dissemination of the results of research already completed. Nothing could be further from the truth. The agricultural research establishments—both privately supported and public—of America and other industrialized nations have focused on large-scale, single crop or livestock operations and labor-saving technologies that are intensive in capital, management and resources. The changing cost of energy, however, is undermining all our previous assumptions about the costs and feasibility of increasing agricultural production and much of the technology of agricultural production as well. No longer can we plan research programs patterned after the conventional ones or those of the past.

We must now develop more diversified resource-conserving technologies for agricultural production. These technologies must maximize output for a given set of inputs, optimize labor utilization and minimize capital costs for development. To be useful, they must improve the economic conditions of farmers. Nations with predominantly small farms must find ways of transmitting new technology to many farmers. This places a great responsibility not only on research programs but also on extension and education.

Environmental issues will become more important as more land, water, fertilizers and pesticides are diverted to food production to force higher productivity. We can expect greater use of chemicals as new technologies are applied. The greatest potential for increases in food production is in the developing countries, and it is in just these countries—mostly tropical and semitropical—that fertilizer needs are greatest and pest problems most acute.

Conflicts in the use of land and water resources for food, feed or fuel production will continue as resource contraints intensify.[10] Toxicities from airborne materials and projected climate changes from fossil fuel emissions will direct attention to the production and use of renewable resources.

Food production is the chief user of our land and water resources. Toxic chemicals in the environment, some of them pesticides and fertilizers used for food production, have been declared hazards to human health and wellbeing. Debates will continue on issues of food safety, deleterious effects of chemicals on fish and wildlife and their habitats, endangered species and carcinogenicity. Although some people have tried, no one has yet clarified what an environmentally sustainable set of agricultural production technologies might be. We must address this issue with more than just debates that result in polarization. This will require a substantial research investment.

Recent history is filled with apocalyptic prophecies of world hunger, famine and starvation. The recently released report of the Presidential

Commission on World Hunger, for example, implies that the food production situation is worsening, and that we are farther from the goal of reducing hunger and malnutrition than we were in 1974. The report presents little evidence, however, to support this statement.[11] Other equally dismal reports ignore prospective scientific discoveries and remain skeptical about major breakthroughs in production. The unfortunate consequence of this pessimism is that without hope there may be little action. Far from achieving scientific or biological limits, however, scientists have only begun to explore the capabilities for increasing food production. Basic and applied research can stimulate future governmental and private sector efforts to increase the stability of production and expand food supplies.[12]

Leadership for the resolution of food production problems through research and technology will continue to reside with the United States. The United States now produces, consumes and exports more food than any nation in all of history. Sixty-one percent of the grain that crossed international borders in 1979 was grown in the United States. Agricultural exports for 1980 approximated $41 billion and offset more than three-fifths of the cost of imported oil. Serious questions have been raised about whether the high U.S. agricultural production can be sustained, especially with the current massive resource inputs. The issue is whether continued, abundant, low-cost foodstuffs can be provided.

U.S. Agriculture and the World Food Situation

With approximately 6.2 billion people on the earth by the year 2000, national strategies must meet increasing demands for improved nutrition and more animal protein, keep food prices reasonable for everyone and lessen tensions among nations. The United States, with its vast human and natural resources, occupies a unique position. Other nations no longer take U.S. supremacy in food and agriculture for granted; yet, they continue to come to our doors in search of new food-producing technologies. The United States cannot and should not plan as a long-term policy to be the breadbasket of the world. This would require an exploitation of land, water, mineral and energy resources that neither we nor the rest of the world can afford for long, if for no other reason than that the price for increasing inputs will likely become prohibitive. The United States, more than any other nation, has already used up its geologic endowment.[13] Because long-distance, massive food transport is energy intensive, it cannot be viewed as a viable long-term alternative to producing food closer to the people who consume it. Several developing countries already have "pockets of success" that employ adaptive sets and combinations of western and domestic technologies.

Great care and restraint should be exercised in using food as a strategic resource. The effects are not always predictable, humane or effective. As the recent grain embargo attempt with the Soviets shows, this kind of strategic use of food penalizes primarily — and perhaps, only — the poor and the farmers. Nevertheless, adequate food supplies can alleviate unrest and tensions among nations and peoples. The U.S. food system faces both domestic and foreign demands that are largely interdependent. Both the balance of payments and exchange of raw materials for value-added goods among the United States and industrial and nonindustrial countries are crucial to the economies of all. Foreign demand on the U.S. food system comes from two different sources: developed or industrialized nations and less developed countries. Food exports from the United States are now going primarily to the industrialized nations and serve mainly to increase the availability of dietary animal protein. In the process, the less developed countries are being largely ignored. It is not likely that this situation will change quickly. Such a global dichotomy will persist.

The objective of U.S. agriculture in the context of the world food situation should be to continue:

- providing a dependable, adequate, safe and nutritious food supply for its domestic needs;

- assisting both industrialized and developing nations, through food exports;

- sustaining a livable environment.

Humanitarian considerations, alleviation of stresses among nations, marketing of surpluses, achieving a balance of payments and needed exchange of materials dictate these objectives. To achieve them, a reassessment of investments in U.S. food and agricultural production research and educational programs, which have progressively eroded since the late 1960s, will be required. The situation has become even more critical since 1977, when the Department of Agriculture (USDA) began elevating consumer and nutrition concerns (food safety, quality, nutritional content) to the highest priority, while deemphasizing food production and marketing research.[14]

The U.S. Agricultural Research System

Food and agricultural research is managed differently in the United States than in other countries. State governments, responding to the aggressive actions of research administrators and scientists at universities, are largely responsible for food and agricultural research. It is the state governments, not the federal government, that provide the bulk (approximately two-thirds) of the money and human resources, establish their own directions, set their own priorities, develop the most innovative approaches on research frontiers and take the initiative in sponsoring foreign agricultural development programs. This stands in marked contrast to research conducted in the defense, space, health, energy and regulatory areas, which is managed largely by federal agencies.

Although there has been a long-standing partnership between the Department of Agriculture and the agricultural experiment stations of the states, that bond is slowly being eroded by a progressive subordination and attrition of cooperative research within the federal system. (One needs only to observe the offices they occupy.) Research has not been a major mission within the Department of Agriculture, although some progress has been made in recent years. More and more financial responsibility for food research is falling upon state governments.

At the same time, the federal agricultural research system is rapidly disintegrating. Few vacancies are being filled. Forty percent of USDA career scientists were due to become eligible for retirement between 1977 and 1982; those who leave are not being replaced.[15] The average age of career scientists in the federal agricultural research system is now 49 and increasing by a third of a year per year. There are progressively fewer young scientists. The system also has been subject to constant personnel attrition, the scientific force having been reduced from 3,300 to 2,850 since the mid-1970s. Meanwhile, new waves of interest and concern—food safety, environmental problems, regulatory contraints, human nutrition, excessive reporting—have been imposed on the research system. Limitations on travel to professional meetings and on funds for operations, imposed by a budget that must allot up to 90 percent of the total financial resources to salaries, provide little incentive for bright young scientists to enter the system.

The Agency for International Development (AID) suffers a similar shortage of agricultural scientists, with only about 300 full-time agricultural positions now remaining.[16] This is shocking if we consider that AID's fiscal year 1979 budget allocated $669 million to agricultural development and nutrition. Thus, only 10 percent of AID's staff is professionally competent to handle agriculture, which is 55 percent of its budget.

Since the early 1970s the purchasing power of federal support of agricultural research has been declining at the rate of 2 percent per year. Final outlays for agricultural research and education in 1980 were 0.3 percent less than the previous year, compared with an average 7 percent gain in federal funding for other types of research and development. Less than 2.3 percent of the total federal research budget of approximately $30 billion was directed to food and agricultural research in 1980; yet global outlays for total research and development for agriculture have averaged 3 percent.[17] We are letting our own national agricultural research system erode while other nations develop theirs. The Congress should intervene immediately to correct this situation.

The United States and the world continue to underestimate and demean the importance of investments in agricultural and food research. Viewed as an investment, with annual returns of 50 percent or more, agricultural research does not receive adequate support in the United States. Two causes have been suggested: First, the benefits of agricultural research spill across countries, states and regions to those who do not pay for it; and, second, benefits to consumers often are not apparent to them.[18]

We should continue to encourage parallel efforts between state and federal support of agricultural research. A decentralized system that addresses the needs of individual states will more than compensate for apparent duplication of effort. Any centralized system designed to achieve maximum coordination among states will only neglect specific regional and state problems and will come at a high price.

Funding of food and agricultural research must include expenses for maintaining and replacing research tools, even when they are not currently being used. These tools include: flocks, herds, barns, feed, milking parlors, machinery, field stations, land, orchards, crops, irrigation equipment and greenhouses. Much of the formula ("Hatch") money traditionally allocated for agricultural research goes into the maintenance of this kind of equipment. As a result, critics repeatedly allege that agricultural research is inefficient compared to the competitive grant programs administered by the National Science Foundation, National Institutes of Health, or—under new legislation—by the USDA. Even though indirect charges are included, these competitive grants do not pay the ever-rising maintenance and replacement costs for machinery, cattle, orchards, crops, land, water, labor or energy. University business offices, however, have seen to it that overhead charges from competitive grant funding pay for on-campus bookkeeping offices, lights, heat and water. This means that agricultural research in universities requires supplementary funding to survive.

Agricultural research in the state agricultural experiment stations is slowing down, not only because prices rise while federal support falls, but because facilities (laboratories, greenhouses, barns and equipment) are woefully inadequate and outmoded. In addition, facilities remain cramped, because student loads have increased threefold since 1968. Yet little federal support has been provided for renovation and improvement of facilities since the mid-1960s, and none since 1970.

Except for maintenance, as outlined above, requests for across-the-board increases for all agricultural research disciplines are no longer convincing.

The message, however, is clear: both competitive grant and formula funding of food and agricultural research should go up, but priorities have to be set for not only the amounts but the kinds of research to be pursued.

Food Production Research Priorities

The food crisis of 1973-1975 and the oil embargo have created new priorities for food and agricultural research. Future research will focus on ways of controlling the biological processes that limit the productivity of economically important food crops and food animals. Research goals will also include more effective use and management of resources and other production inputs.

The national or international working conference model has been an effective means of establishing priorities in agricultural research. The best scientific talent with a range of interdisciplinary skills is recruited. Commissioned papers on specified topics are prepared and distributed to prospective participants in specified working groups in advance of the conference. After a week's revision and further development during the conference, the results are edited and published as proceeedings. An executive summary sets forth the priorities.

Assessments of research priorities for the plant and animal sciences have been elaborated in an international conference on crop productivity,[19] several reports of the U.S. National Research Council—National Academy of Sciences,[20] a national conference, "Animal Agriculture—Meeting Human Needs for the 21st Century"[21] and by the Office of Technology Assessment of the U.S. Congress.[22] The World Food Conference of 1976 also outlined research priorities.[23] A working conference, sponsored by the Agricultural Research Policy Advisory Committee of the state-federal system, established research priorities by a ballot system from a large number of participants.[24] The International Conference on Agricultural Production—Research and Development Strategies for the 1980s issued recommendations for research and development in biological resources, soils, water and energy.[25] Within the past few years, public and private agricultural and food research centers, from provincial to international levels, have reassessed and identified research priorities. These centers have sponsored long-range planning seminars on the major issues and trends of agricultural science and technology. From all of these efforts, a surprising unanimity has emerged.

Biological Research and Food Production

Through research, scientists could develop technologies that would result in stable food production at high levels. These technologies would enhance rather than diminish the earth's resources. They would not pollute the environment or use large amounts of capital, management or nonrenewable resources. They would be scale neutral. Development of these technologies is of the highest priority and can be accomplished through biological research. Through biological research, we can take steps toward enabling plants and food animals to use present environmental resources more effectively. Through biological research we can achieve:

(1) greater photosynthetic efficiency;

(2) improved biological nitrogen fixation;

(3) genetic improvements;

(4) more resistance to competing biological systems (weeds, insects, diseases);

(5) more efficient nutrient and water uptake and utilization, and fewer losses from nitrification and denitrification of nitrogen fertilizer applied in crop production;

(6) alleviation of climate and environmental stresses (unfavorable temperatures, soil moisture and mineral stresses in problem soils);

(7) better understanding of hormonal systems and their regulation.

These technologies may release food production from dependence on increasingly scarce fossil fuels.

Efforts to identify these important research areas have not resulted in expanded research support, but they have prompted changes in organization, administration and funding of agricultural research at the federal level. A notable example has come with the initiation of the competitive grant program administered by USDA. This infant program, which supports the first four of the research areas listed above, was initiated in 1978. Announcement of the $14 million program brought in more than 1,100 research proposals involving funding requests for more than $200 million. Available funds could support only half the proposals rated as excellent. Similar situations existed in 1979 and 1980.

All of this has revealed one important fact. There is enormous talent waiting to be recruited for viable research programs related to the biological processes that control food production. Nevertheless, during the young life of the competitive grant program, now in its fourth year, Congress has consistently limited funding of the program to essentially the same level, denying the program even those increases needed to offset the effects of inflation. The minimal increase allowed has been eaten up in administrative costs. The available human resources revealed by the number of applicants with excellent project proposals herald an opportunity for this nation to reassert the world leadership that it has abdicated in the area of food research. A policy that severely limits funding, however, denies to agricultural and food research programs the talents of some of the very best scientists in the nation. This cannot be reconciled with our true national interest. It is time we opened the door of agricultural and food research to the nation's scientific expertise, including that possessed by the private sector.

The benefits of support for the competitive grant program would be reaped by the developing countries as well as by the United States. Developing nations can share in the benefits of the new technologies we have already discussed — improved plant and animal genetics, increased photosynthetic efficiency and nitrogen fixation, as well as protection against insects, diseases, weeds and adverse environments. These technologies can free developing nations as well as the United States from an ill-advised dependency on fossil fuels for food production. The research necessary to create these new technologies is adaptable to local conditions and is relatively inexpensive.[26]

Benefits from such research could be multiplied many times, and advances made by one nation could be shared by others. Genetic pools could be assembled, for example, so that nations could share information about known favorable components for disease resistance, environmental stress tolerance, a superior "harvest index" (the portion of the plant used for food) and acceptable culinary characteristics that can be adapted quickly to local needs and conditions.

A particularly significant, yet neglected, biological research area is the alleviation of climatic and environmental stresses. The report of the Steering Committee for the National Academy of Sciences World Food and Nutrition Study, issued in 1977, deemed this as important as improved photosynthesis and biological nitrogen fixation, genetic manipulation and protection against pests. The effects of climate and weather remain the most significant determinants in food production and account, more than any other inputs, for instabilities in production from year to year and from nation to nation.

Stability of production is as important as the magnitude of production itself. Climate is probably a more significant determinant of food production than are pests. The droughts of 1974 and 1980, for example, caused far greater losses of U.S. agricultural production than the blight that destroyed 15 percent (or about 700 million bushels) of the U.S. corn crop in 1970. In 1974, production plummented 20 percent for corn, wheat and soybeans as a result of drought. In 1980, corn production fell 17 percent from 1979, or 1.3 billion bushels; grain sorghum 32 percent; feed grains 18 percent; soybeans 22 percent; cotton 23 percent; and peanuts 43 percent.

Climatic stresses on world grain production were particularly significant

Figure 8.2. Patterns of World Grain Production in Millions of Metric Tons (MMT).

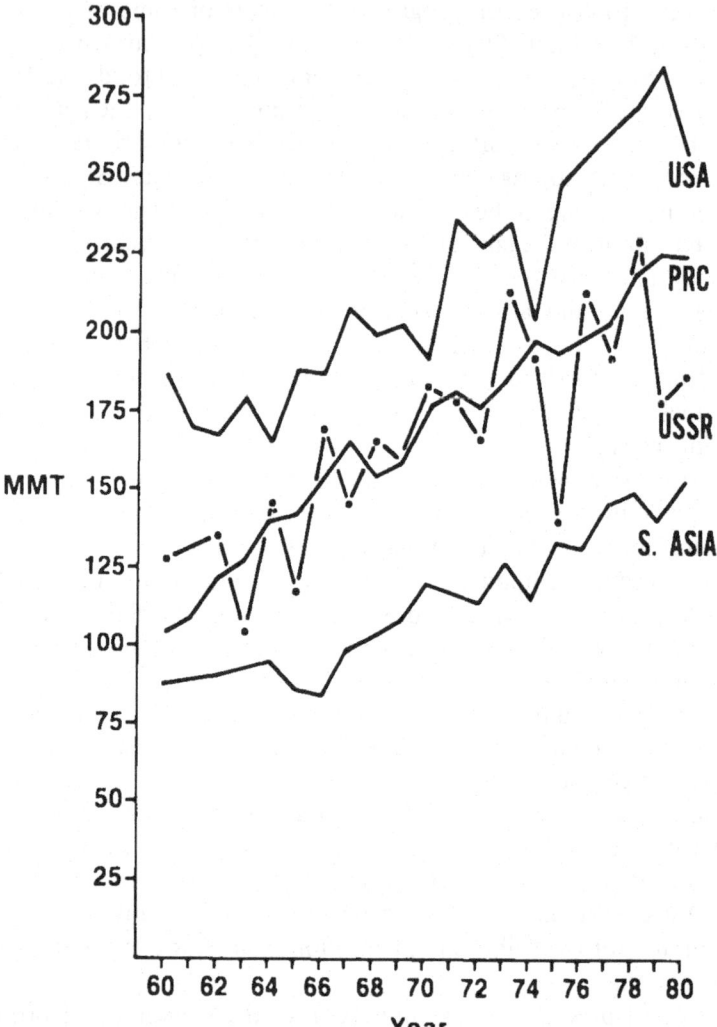

Note: For the United States, People's Republic of China, Soviet Union, and South Asia from 1960 to 1980. Perturbations in the upward trends are primarily a result of climatic and weather events. The extreme variations in the USSR result from millions of hectares of land that are marginally cold or dry. The relative stability of production in China and South Asia results from a high percentage of crop irrigation. Climatic stresses during 1980 will result in a severe depletion of grain stocks and storage reserves. *Source:* U.S. Department of Agriculture.

in 1980. That year, the composite index of crop yields in the United States dropped 20 percent because of drought and high temperature. Grain production fell off in the People's Republic of China because of floods in the south and drought in the north. The Soviet Union witnessed its second disastrous year in a row because of marginally cold and dry growing conditions and adverse weather during harvest. Only in South Asia did production rise slightly above previous highs (Figure 8.2). These statistics suggest that world and U.S. grain stocks will be reduced, relative to utilization, to the lowest levels encountered in two decades.

A substantial research effort aimed at improving the resistance of crops to stresses caused by interrannual climate variations is badly needed. Although potential problems of long-term climate change—for example, from increasing levels of atmospheric CO_2—have received considerable attention in recent years, problems of interannual climate variation have been largely overlooked. Research directed at achieving greater production stability through genetic improvement and crop and livestock management is of high priority. The USDA's competitive grant program must be expanded to accommodate this kind of research, and the Congress should respond accordingly.

Some new research initiatives are also called for in animal agriculture and its products. Three-fourths of the dietary protein, one-third of the energy and most of the calcium and phosphorus in the U.S. diet come from animal products. Food animals, such as ruminants, are living, mobile protein factories that may survive and flourish on forages that are indigestible for people. Food animals like swine and poultry use residues and by-products from food processing and from the polycultures of lakes and ponds that otherwise would be wasted. The world food reserves in livestock exceed those of grain and are far better distributed.

The feeding of livestock in other nations is the catalyst for much of the current U.S. world grain trade. Livestock, not people, consume most of the corn, wheat and soybeans the United States ships abroad. People of all nations, developed and less developed, are striving for more dietary animal protein. Research in animal agriculture should focus on resource conservation, greater reproductive efficiency and basic studies on protein synthesis that would result in less fat and more protein in the final product.

Human Resources for Agricultural Research

Consideration must be given to human resource needs for food and agricultural research and technology. We have already mentioned the serious loss of career scientists in the federal agricultural research system. Mention is often made of the lack of social science inputs into the nation's agricultural research program. An almost exclusive responsibility for agricultural research training programs in the United States resides with the land grant universities, a few additional state universities, the colleges of 1890 and the Tuskegee Institute. These institutions train scientists for research in state agricultural experiment stations, the USDA's research programs, cooperative state-federal programs, the private sector, the foundations and the international agricultural research centers. The 15 top land grant universities, each with enrollments of 100 or more foreign graduate students, have now produced 20,000 alumni helping to serve agricultural research and educational needs in developing countries. These alumni are one of the greatest resources this nation has cultivated for contributing to the future role of U.S. agriculture in the context of the world food situation. Most of the 10,000 to 12,000 U.S. agricultural scientists who receive public support, and an even greater number from the industrial sector, plus many of the more than 600 senior scientists in the international agricultural research centers, are also alumni of the U.S. land grant system.

The human resource base for scientific support of food research in the United States now has fallen behind the Soviet Union and the People's Republic of China.[27] In the United States, shortages of trained scientists are emerging in agricultural economics, agronomy, engineering and animal agriculture. The United States can expect increasing demands, both at home and abroad, for training and aiding agricultural scientists. These demands call for a review of the entire training program and raise serious questions about where international agriculturists will come from.[28] The National Science Foundation is responsible for the health of science in the nation. It must reassess its role in supporting the biological, physical and social sciences in research on food production and distribution and in supporting foreign graduate students for training programs in agriculture and food production.

International Agricultural Research Centers

Globally, the most successful agricultural research establishments are the international centers. They have undertaken innovative research projects for enhanced food production and other aspects of agriculture and have prospered. Annual funding for these centers (now numbering 13) has gone from $10 million in 1969 to more than $125 million in 1980, with $250 million projected for 1984. The United States continues to contribute about 25 percent of their total budget which, along with other sources of income, is administered by the Consultative Group on International Agricultural Research with advice and scientific input from its Technical Advisory Committee.

Recent congressional and administrative proposals are aimed at consolidating all foreign research and technology activities, along with many other programs, under a new International Development Cooperation Agency (IDCA). This agency would include the Institute for Scientific and Technological Cooperation (ISTC), which would have a strong input from U.S. colleges and universities. ISTC would have responsibility for science and technology efforts in developing countries relating to food and agriculture. The current Collaborative Research Support Programs (CRSPs) would then fall under the administration of ISTC.

An appropriate constituency has not yet been developed either in Congress or in the nation to support ISTC or expanded international agricultural development programs.[29] Approximately half ($50 million) of the proposed 1981 budget of ISTC would have focused on food, nutrition and agricultural programs in other nations, with primary focus on developing countries. Congressional opinion, however, does not support federal programs for food and agricultural research directed toward the needs of other nations.[30]

This lack of support is puzzling in view of the benefits we ourselves can derive from international involvement. Most of our major food crops and breeds of livestock and much of our technology have been derived from other countries:

- dwarf — high yielding — varieties of wheat and rice from Japan,

- soybeans from China,

- insect-resistant wheat from Russia,

- new genetic resources for third-generation hybrid corn from South America,

- high Vitamin A sorghum from West Africa,

- high protein, high lysine wheat from Nepal,

- cattle more tolerant of heat, parasites and insects from Africa and Asia.

These are by no means the only benefits to be derived from other countries. The production of Zebu (Brahma) cattle from the Asian subcontinent has created an entirely new beef industry in the higher temperature regions of the southern United States in less than 20 years. The most advanced genetic material for dwarf hybrid sunflowers resides in the Soviet Union. Future collaboration with the People's Republic of China is expected to make available vast genetic resources in swine breeding, cereal grains, oil crops and many yet undeveloped fruits and vegetables. The Chinese also have much to offer in Azolla culture (green manuring through biological nitrogen fixation) and hybrid rice production.

Congress should review carefully what is emerging from the CRSPs that are administered by the Board on International Food and Agricultural Development. Teams of U.S. economists and other social scientists are participating in interdisciplinary and interuniversity programs committed to research design and implementation. They are collaborating on both basic and applied research with similar groups from developing countries. The major objective of such programs is more effective resolution of a wide variety of staple food problems, including production, utilization and the sociocultural impacts resulting from them. It is implicit in the CRSPs that research will emphasize technologies that do not exploit resources, pollute the environment or depend on large energy inputs. Biological solutions will be emphasized, wherever possible, over reliance on costly and possibly polluting agricultural chemicals. This kind of research may be significant to U.S. agriculture as we move toward a more resource-conserving mode.

It is expected that the benefits of the research will have a global impact incorporating U.S. interests, because we have been active in planning strategy from the outset. Established efforts involving collaboration among U.S. institutions, international research centers and commodity networks and national research centers include programs on field beans and cowpeas, sorghum-millet, small ruminants, integrated pest management and aquaculture. One of the most advanced of the CRSP efforts is the field beans and cowpeas program, which has locations outside the United States, including 12 research institutions in Latin America and east Africa and 2 international agricultural research centers. The program involves 10 U.S. universities and brings together many disciplines, including agronomy, botany, genetics, plant pathology, entomology, food science, human nutrition, medicine and social science. Managed by Michigan State University and guided thus far by a sociologist and a plant breeder, the program's initial contract calls for $16.7 million for a five-year period.

Challenges ahead for the CRSPs will be to seek funding at the federal level for up to 50 percent of the U.S. investigators' time. The current support level of about 10 percent is disproportionate for the managing institution and cannot survive. There should be an effort to train counterpart scientists in the developing countries and to promote regional centers for training intermediate level technicians and extension personnel, both men and women. CRSPs can help bridge the gap that now exists between international agricultural research centers and national programs. They can help develop and hold together global research teams on specific problems. Through these collaborative efforts, it is hoped that recognition will be given to inputs from cultures where the solutions are to be used. Many peasant farming practices are also worthy of research, and some may have useful applications for U.S. agriculture.

The role of the international agricultural research centers is under constant review[31] and should go beyond inputs from the Consultative Group on International Agricultural Research. Major early breakthroughs, such as occurred with dwarf types of wheat and rice, characterized their early development. Gaps now exist between the international centers and national agricultural research centers. To bridge these gaps will require closer collaboration in the future between the two. The international agricultural research centers are not yet truly international, because much of the world is not a part of the networks either as contributors or as recipients. The international agricultural research centers also exhibit and exercise a degree of research affluence (higher salaries, benefits, equipment, supporting personnel, travel opportunities) not typical of the countries in which they are located. It is unlikely that any of these problems will be overcome soon.

It is recognized that the international agricultural research centers have, in some instances, upgraded national agricultural research centers. The National Institute for Agricultural Research in Mexico is a good example.

Emphasis in the future should be on the increased support and development of national agricultural research centers. Most food production problems are regional, and solutions must be localized. There is a movement toward this with the establishment of the International Agricultural Development Service and the International Service for National Agricultural Research, the most recent member of the international agricultural research network.

It is further recommended that the Consultative Group on International Agricultural Research take the initiative in establishing two international research centers for forestry. One center should be located in the tropics, with Brazil, Indonesia or Africa as possible sites; the other, in the temperate zones of either North America or China. Attention should be given to enhancement of forest productivity through genetic improvement and management. Special emphasis would be given to biomass as a renewable energy resource, reforestation and control of soil erosion, trees and their products as food resources and the technologies of agriforestry, utilizing species that have biological nitrogen fixation capabilities.

Conclusion

The United States cannot indefinitely serve as the breadbasket of the world. Food production and its delivery, along with fossil fuel energy, will become increasingly expensive, and at times both food and energy will be

Table 8.1. Pockets of Successful Production in Developing World Agriculture.

Projects	Accomplishments
Grain Production in India's Punjab	A 3-fold increase in grain production in 10 years from 1965 to 1975
Rice Production in Colombia	Yields rose from 1.8 to 4.4 tons/hectare from 1965 to 1975
Wheat Production in Turkey	Increase in production from 7 to 17 million tons from 1961 to 1977
Hybrid Rice in China	30 to 50 percent yield increase (labor intensive high yielding technology)
Hybrid Cotton In India	Yields doubled (labor intensive high yielding technology)
The White (Milk) Revolution in the Gujarat State of India	Daily cash income, improved nutrition, labor intensive technology for 300,000 small farms
The Puebla (Maize) Project in Mexico	Yields increased by 30 percent from 1968 to 1972
The Comilla Project of East Pakistan (Bangladesh)	Rice yields and incomes of farmers doubled from 1963 to 1970
The "Masagana 99" Project in the Philippines	Rice yields increased by 36 percent in 3 years from 1973 to 1976
Maize in Kenya	Hybrids and fertilizer and management increased yields 4.8 tons/hectare
Hybrid Maize in the U.S.	3.5 fold increase in yield from 1940 to 1979

Source: Author, and S. Wortman and R.W. Cummings, Jr., To Feed This World: The Challenge and the Strategy (Baltimore: Johns Hopkins University Press, 1978), pp. 186-230.

scarce. Agricultural development must precede economic development. Ultimately, the answer will dictate that food be produced closer to the people who consume it. To this end, there are notable examples of successful food-producing systems in the agriculturally developing world. Some of them are summarized in Table 8.1. The technological, social, economic and resource ingredients that have gone into these pockets of success should be identified and shared with other nations where their adoption could prove equally fruitful.

Notes

1. The author in the preparation of this paper has drawn heavily from his previous reports. These include "Maximum Production Capacity of Feed Crops," *BioScience*, vol. 24 (1974), pp. 216–44; "Food Production: Technology and the Resource Base," *Science*, vol. 188 (1975), pp. 579–84; "Increased Crop Yields and Livestock Productivity," in *World Food Prospects and Agricultural Potential*, Marilyn Chou, David P. Harmon, Jr., Herman Kahn and Sylvan H. Wittwer (New York: Praeger Publishers, 1977), pp. 66–135; "Assessment of Technology in Food Production," in *Renewable Resource Management for Forestry and Agriculture*, James S. Bethel and Martin A. Massengale, eds. (Seattle: University of Washington Press, 1978), pp. 35–56; "Production Potential of Crop Plants," in *Crop Physiology*, U. S. Gupta, ed. (New Delhi: Oxford and IBH Publishing Co., 1978), pp. 334–73; "The Shape of Things to Come," in *Biology of Crop Productivity*, Peter Carlson, ed. (New York: Academic Press, 1980), pp. 413–59; "Food Production Prospects: Technology and Resource Options," in *The Politics of Food*, D. Gale Johnson, ed. (Chicago: University of Chicago Press, 1979); "Priorities of U.S. Food Research and Management," a position paper, Presidential Commission on World Hunger, vol. 3 (Washington, D.C., 1980); "Agriculture for the 21st Century," the Coromandel Lecture, New Delhi, India, 4 September 1979; "Future Trends in Agriculture Technology and Management," *Long-Range Environmental Outlook*, Environmental Studies Board of the Commission on Natural Resources, National Research Council (Washington, D.C.: National Academy of Sciences, 1980), pp. 64–107; "Agriculture in the 21st Century," *Proceedings of the Agricultural Sector Symposia* (Washington, D.C.: World Bank, n.d.), pp. 450–95; "The Role of Science in Future Food Production Increases," paper presented at the International Wheat and Maize Development Center Long-Range Planning Conference, 9–12 April 1980; "Advances in Protected Environments for Plant Growth," in *Advances in Food-Producing Systems for Arid and Semiarid Lands*, Jamal T. Mannassah and Ernest J. Briskey, eds. (New York: Academic Press, 1981), pp. 679–716; "Future Challenges and Opportunities for Agricultural and Forestry Research," in *Future Sources of Organic Raw Materials*, I. Chemrann, L. E. St. Pierre and G. R. Brown, eds. (New York: Pergamon Press, 1980), pp. 401–12; and "Food for the 21st Century," keynote address in *Animal Agriculture: Research to Meet Human Needs for the 21st Century*, Wilson G. Pond, Robert A. Merkel, Lon D. McGilliard, and V. James Rhodes, eds. (Boulder, Colo.: Westview Press, 1980), pp. 331–44.

2. "The Future of American Agriculture as a Strategic Resource," *Proceedings of a Conservation Foundation Conference* (Washington, D.C.: Conservation Foundation, 1980).

3. *Presidential Commission on World Hunger—Final Report* (Washington, D.C.: 1980); and C. R. Wharton, Jr., "Food—The Hidden Crisis," *Science*, vol. 208 (27 June 1980), p. 1415.

212 Sylvan H. Wittwer

4. World Food and Nutrition Study, *Enhancement of Food Production for the United States* (Washington, D.C.: National Research Council/National Academy of Sciences, 1975); and *The Potential Contributions of Research* (Washington, D.C.: National Research Council/National Academy of Sciences, 1977).

5. Sylvan H. Wittwer, "The Shape of Things to Come," in *Biology of Crop Productivity*, Peter Carlson, ed. (New York: Academic Press, 1980); *Global Food Assessment*, Foreign Agricultural Economic Report No. 159 (Washington, D.C.: U.S. Department of Agriculture, 1980).

6. Sterling Wortman and R. W. Cummings, Jr., *To Feed This World* (Baltimore: Johns Hopkins University Press, 1978), p. 406.

7. *Agriculture Toward 2000* (Rome: Food and Agriculture Organization of the United Nations, 1979).

8. Council on Environmental Quality and the Department of State, *The Global 2000 Report to the President: Entering the Twenty-First Century* (Washington, D.C., 1980); and A. M. Tang and B. Stone, *Food Production in the People's Republic of China*, Research Report 15 (Washington, D.C.: International Food Policy Research Institute, 1980).

9. Vernon W. Ruttan, "Induced Innovation and Agricultural Development," *Food Policy*, vol. 2, no. 3 (August 1977), pp. 196–216; and Yujiro Hayami and Vernon W. Ruttan, *Agricultural Development: An International Perspective* (Baltimore: Johns Hopkins University Press, 1971).

10. L. R. Brown, *Food or Fuel: New Competition for the World's Cropland*, Worldwatch Paper 35 (Washington, D.C.: Worldwatch Institute, 1980); R. Meekhof, M. Gill and W. Tyner, "Gasohol: Prospects and Implications," in *Agricultural Economic Report No. 458* (Washington, D.C.: Economics, Statistics, and Cooperative Service, U.S. Department of Agriculture, 1980).

11. *Presidential Commission on World Hunger—Final Report;* and D. Gale Johnson, *The World Food Situation: Development during the 1970s and Prospects for the 1980s*, Paper No. 80 (Chicago: Office of Agricultural Economics Research, University of Chicago, 1980), p. 10.

12. For further development of these concepts, see Chou et al., *World Food Prospects;* Marilyn Chou and David P. Harmon, Jr., eds., *Critical Food Issues of the Eighties* (New York: Pergamon Press, 1979); "The Dynamics of Natural Resources and Their Impact on Resource Data," paper presented at the Conference on Systems Aspects of Energy and Mineral Resources, Laxenburg, Austria, 9–14 August 1979, sponsored by the International Institute for Applied Systems Analysis and the Resource System Institute, East-West Center, Honolulu; Keith O. Campbell, *Food for the Future* (Lincoln: University of Nebraska Press, 1979); Theodore W. Schultz, *The Economics of Research and Agricultural Productivity* (New York: International Agricultural Development Service, 1979); Johnson, *The World Food Situation;* Pinstrup P. Anderson, *The Impact of Technological Change in Agriculture on Productivity, Resource Use and the Environment: Towards an Approach for Ex Ante Assessment*, Working Paper WP-79-08 (Laxenburg, Austria: International Institute for Applied Systems Analysis, 1979); E. Ebeling, *The Fruited Plain* (Berkeley: University of California Press, 1979); N. S. Scrimshaw and L. Taylor, "Food," *Scientific American*, vol. 243 (1980), pp. 78–88.

13. See Chapter 10, by William A. Vogely, in this volume.

14. The magnitude of the task of focusing on the policy implications for U.S. science and technology, along with interrelationships among situations for U.S. agriculture in the context of the world food situation, during the next five years forced a limitation in the coverage of this report. The entire food system from production through handling, processing, marketing and consumption is not addressed. Nor is that of aquaculture. There is the recognition, however, that food losses beyond the farm gate are significant and that a reduction in these losses could result in a 10 to 15 percent increase in food supply without bringing new land into production; see *Food Science Research for Improving the Utilization, Processing and Nutritional Value of Food Products* (Washington, D.C.: National Science Foundation, 1975). The focus of this report is on science and technology options, with priorities, for food production.

15. See the report requested by the late Senator Hubert Humphrey entitled *Management of Agricultural Research: Need and Opportunities for Improvement* (Washington, D.C.: General Accounting Office, 23 August 1977).

16. J. Knoll, *BIFAD Study of AID's Professional Agricultural Manpower,* Board for International Food and Agricultural Development (Washington, D.C.: International Development Cooperation Agency/Agency for International Development, 1979); R. O. Olson, *Staffing Requirements for Agriculture/Rural Development Programs* (Washington, D.C.: Agency for International Development, 1980).

17. For further details see: "Research and Development Special Analysis," *The Budget of the U.S. Government for Fiscal 1980* (Washington, D.C., 1979), p. 304; Colin Norman, *Knowledge and Power: The Global Research and Development Budget,* Worldwatch Paper 31 (Washington, D.C.: Worldwatch, 1979); and A. R. Bertrand, "Food and Agricultural Research Agenda for the 80s," and the response by Don Paarlberg, paper presented at the USDA Agricultural Outlook Conference, Washington, D.C., 19 November 1980.

18. R. E. Evenson, P. E. Waggoner and V. W. Ruttan, "Economic Benefits from Research: An Example from Agriculture," *Science,* vol. 205 (14 September 1979), pp. 1101-7.

19. "Crop Productivity—Research Imperatives," in *Proceedings of an International Conference,* A.W.A. Brown, T. C. Byerly, M. Gibbs and A. San Pietro, eds. (East Lansing, Mich.: Michigan Agricultural Experiment Station, 1976).

20. World Food and Nutrition Study, *The Potential Contributions of Research.*

21. *Animal Agriculture—Meeting Human Needs for the 21st Century,* Proceedings of a National Conference, 4-9 May 1980, Boyne Mountain, Mich. (Boulder, Colo.: Westview Press, 1980).

22. U.S. Congress, Office of Technology Assessment, *Organizing and Financing Basic Research to Increase Food Production* (Washington, D.C.: 1977).

23. *Proceedings of the World Food Conference of 1976* (Ames: Iowa State University Press, 1976).

24. *Research to Meet U.S. and World Food Needs,* report of a working conference (Agricultural Research Policy Advisory Committee, Kansas City, Mo., 9-11 July 1975).

25. *Agricultural Production: Research and Development Strategies for the 1980s* (New York: Rockefeller Foundation, 1980).

26. Roger Revelle, "Energy Dilemmas in Asia: The Needs for Research and

Development," *Science,* vol. 209 (4 July 1980), pp. 164–74; and "Biological Research and the Third World Countries" (editorial) *BioScience,* vol. 3, no. 1 (1980), p. 727.

27. Sylvan H. Wittwer, "U.S. and Soviet Agricultural Research Agendas," *Science,* vol. 208 (18 April 1980), p. 245; and "Agricultural Research: Some Comparisons of the Soviet, Chinese, and the United States Systems," *National Forum,* vol. 61 (Winter 1981), pp. 20–21.

28. Clifton R. Wharton, Jr., "Tomorrow's Development Professionals: Where Will the Future Come From?" banquet address, American Agricultural Economics Association, Urbana, Ill., 28 July 1980.

29. See Chapter 7, by Charles Weiss, in this volume.

30. Some of the most effective programs have been based on bilateral relationships between U.S. land grant or other universities and foreign institutions. An excellent example is that of Ohio State University and the Punjab Agricultural University in India.

31. Consultative Group on International Agricultural Research, *1980 Report: An Integrative Report* (Washington, D.C.: CGIAR Secretariat, 1980).

9
Trends and Prospects in World Population

Michael S. Teitelbaum

Population trends form the basic substrate of concerns about resources and international comity over the coming decades. The links between population change and the other matters under discussion here often are not single and direct but instead are mediated by a broad array of political, technological and economic factors. The extraordinary force of recent population change, however, should not be minimized. Present trends have no precedent in human experience in terms of current population size, percentage rates and absolute size of increase and ultimate projected population size. Hence, projections into the future, apart from the normal limitations of all projections, represent leaps of faith into population aggregates that are off the scale of experience.

Recent Population Trends

This is not the place for an introduction to demography or for a lengthy review of recent population trends, but a brief summary is in order.[1] Three components of population change are central—fertility, mortality and migration. Each of these components is measurable (with greater or lesser accuracy), and each has distinctive effects upon the characteristics of human populations on a global, national and subnational level.

The postwar period has seen substantial, sometimes dramatic, mortality declines in both developed and developing countries. In developed countries there have also been more erratic changes in fertility patterns, with most showing postwar recovery from the record-low fertility levels of the 1930s (the so-called baby boom in the United States was an extreme example). The postwar recovery was followed almost universally by rapid fertility declines

Michael S. Teitelbaum is a program officer with the Ford Foundation, New York, N.Y.

in the 1960s and 1970s, reaching levels that are now frequently lower even than those of the 1930s.

A different set of patterns prevailed in the developing countries. In some countries, notably China, Taiwan, the Philippines, Tunisia and Colombia, mortality declines have been followed by substantial fertility declines during the 1960s and 1970s from previously very high levels. In other countries, such as Bangladesh, Pakistan and Egypt, fertility has declined little, if at all. In still other countries, such as Kenya, fertility apparently has increased dramatically. The overall result has been acceleration in population growth in much of the developing world, followed by modest recent declines in growth rates in some areas. The experience over the past three decades may be seen in Table 9.1.

The world's population increased by 1.9 billion, or over 75 percent, in the three decades from 1950 to 1980, as Table 9.1 indicates. The annual rate of increase was at a then-record level of 1.77 percent in the 1950–1955 period but actually accelerated sharply in the 10 years to 1960–1965, peaking at 1.99 percent (Table 9.2). This was followed by a slower decline in the rate of increase, to a level of 1.81 percent in 1975–1980 — lower than the peak rate but still higher than the 1950–1955 quinquennium.

Because the higher percentage rates of economic growth, inflation or interest are more familiar than demographic rates, it is important to keep recent population growth rates in a proper demographic perspective. Sustained annual population growth rates of 1.8 to 2.0 percent have never before been seen for the world population as a whole. Because of continuous compounding, they imply a doubling of world population every 35 to 39 years. They fit within a frame in which the theoretical maximum growth rate — assuming excellent mortality conditions and no restraint on fertility — is on the order of 4.0–4.5 percent. (By contrast, inflation and interest rates presumably have no upper bound.)

There is reason for concern that the fertility declines of the past 15 years have been widely misinterpreted as evidence that problems of rapid population growth have been resolved.[2] The achievements in fertility reduction in the 1960s and 1970s were significant indeed but so were the achievements in mortality reduction. As a result, growth rates declined only modestly, from about 2 percent to about 1.8 percent. The problem of high population growth rates has not been resolved in the past decade. Rather the trend has reached a point of inflection. The rate of increase itself is no longer increasing but instead has declined modestly, although remaining at extraordinarily high levels compared with all past human experience. Furthermore, it must be noted that the absolute size of population growth is as important as percentage increase, and here a few numbers illustrate the uniqueness of the past decades. The population of Asia alone in 1980 (2.558 billion), for ex-

Table 9.1. Population (Millions) in the Eight Major Areas of the World, 1950 to 2000.

Year	World	Africa	Latin America	Northern America	East Asia	South Asia	Europe	Oceania	Soviet Union
				Estimates					
1950	2513	219	154	166	673	706	392	13	180
1955	2745	244	187	182	738	775	408	14	196
1960	3027	275	215	199	816	867	425	16	214
1965	3344	311	247	214	899	979	445	18	231
1970	3678	354	283	226	981	1111	460	19	244
1975	4033	406	323	236	1063	1255	474	21	254
1980	4415	469	368	246	1136	1422	484	23	267
				Projections					
1985	4830	545	421	258	1204	1606	492	24	280
1990	5275	630	478	270	1274	1803	501	26	292
1995	5733	726	541	281	1340	2005	510	28	302
2000	6199	828	608	290	1406	2205	520	30	312

Note: Trends are given as they were assessed in 1978. 1980 data are projections from mid-1975 data, but may be viewed as best available estimates for 1980.

Source: W. Parker Mauldin, "Population Trends and Prospects," Science, vol. 209 (4 July 1980), p. 156.

Table 9.2. Average Annual Rate of Increase (Percentage) in the Eight Major Areas of the World.

Year	World	Africa	Latin America	Northern America	East Asia	South Asia	Europe	Oceania	Soviet Union
				Estimates					
1950 to 1955	1.77	2.16	2.72	1.80	1.85	1.86	0.79	2.25	1.71
1955 to 1960	1.95	2.36	2.78	1.78	1.99	2.24	0.84	2.18	1.77
1960 to 1965	1.99	2.49	2.77	1.50	1.94	2.44	0.90	2.09	1.49
1965 to 1970	1.90	2.61	2.67	1.11	1.75	2.52	0.66	1.96	1.09
1970 to 1975	1.84	2.71	2.64	0.87	1.62	2.45	0.61	1.82	0.84
1975 to 1980	1.81	2.91	2.66	0.83	1.32	2.49	0.39	1.47	0.94
				Projections					
1980 to 1985	1.80	2.97	2.65	0.96	1.16	2.44	0.36	1.41	0.94
1985 to 1990	1.76	2.93	2.58	0.91	1.14	2.31	0.35	1.37	0.85
1990 to 1995	1.66	2.81	2.46	0.76	1.01	2.13	0.37	1.30	0.70
1995 to 2000	1.56	2.64	2.34	0.61	0.95	1.91	0.38	1.19	0.64

Note: Trends are given as they were assessed in 1978. 1980 data are projections from mid-1975 data, but may be viewed as best available estimates for 1980.

Source: W. Parker Mauldin, "Population Trends and Prospects," Science, vol. 209 (4 July 1980), p. 156.

ample, is larger than the entire world population was in 1950 (2.513 billion). A population about as large as the combined 1980 populations of Europe and North America (excluding European portions of the Soviet Union) was added to the already existing population of South Asia in only 30 years (an increase of 716 million from 1950 to 1980). Many similar comparisons could be made to illustrate the extraordinary demographic experiences since World World II.

Table 9.2 provides a useful retrospective view of regional growth rates. Africa and South Asia show clear patterns of accelerating growth rates in the 1950s and 1960s, while the other regions were more mixed. Since the 1960s most regions have experienced substantial declines in their rates of population increase. The very large region of South Asia, however, shows no such decline, and the smaller region of Africa shows a rapid and continuing increase in rate of increase right up to the present.

The third population change component – migration – appears to be of large and rapidly increasing magnitude. There have been, over the past decades, increasing movements within the countries from rural to urban areas. In the developing world, urban growth rates are about twice as high as the already high national growth rates. These rapid growth rates suggest unprecedented urban agglomerations appearing over the coming 20 years.

International migration, including legal or illegal, political or economic, temporary or permanent, has also grown rapidly. Both internal and international migration have important, if elusive, implications for issues of resources and international relations. Movement of rural populations from subsistence economies to more energy-intensive urban areas, for example, presumably implies higher per-capita energy needs. Similarly, large migrations from developing countries to developed countries suggest greater overall demand for food, mineral resources and energy.

Population Trends in Prospect

It must be freely admitted that population projections do not predict but, rather, represent the logical implications of assumed future trends in fertility, mortality and migration. Population trends, however, are relatively stable compared with the political and economic, due to a three-part, built-in inertia in demographic change to the year 2000. First, the majority of persons who will be living then are already alive; second, human reproductive behavior changes relatively slowly; and third, high fertility generates a youthful population with strong momentum for continued growth over many decades. As a result, the demographer can predict trends with reasonable accuracy over several decades, although not much beyond. Projections of population change to the year 2000, in contrast to those for

economic or political change, have considerable plausibility, barring unpredictable catastrophes.

The conventionally accepted multinational population projections are
those prepared periodically by the United Nations (U.N.) Population Division; others are available from the World Bank and the U.S. Bureau of the
Census. Given demographic inertia, available projections to the year 2000
are broadly compatible. The medium variant of the widely accepted U.N.
projections is summarized in Table 9.2. It shows a projected gradual decline
in world population growth rates of about one-quarter of 1 percent,
reaching 1.56 percent overall in 1995–2000. At the same time, the rapidly
growing population base means that the numbers of people added each year
will continue to grow—to about 90 million additions per year in 1995–2000
versus about 75 million annually in 1980, despite the projected decline in
growth rate.

The three variants in U.N. projections show a total world population in
2000 of 5.9, 6.2 and 6.5 billion. Most other projections cluster around 6
billion. Again using the medium variant, overall growth of 40 percent (1.78
billion) is projected, with regional increases of 77 percent in Africa (359
million), 65 percent in Latin America (240 million), 55 percent in South
Asia (783 million), 24 percent in East Asia (270 million), 18 percent in North
America (44 million), 17 percent in the Soviet Union (45 million) and 7 percent in Europe (36 million).

The momentum of population growth in the developing countries is likely
to continue, although projections beyond the year 2000 are quite speculative. No one can anticipate the course of fertility change in high fertility
regions of South Asia and Africa, where fertility has not yet begun to
decline. For this reason, and because catastrophes are possible but unpredictable, there is some consensus as to the plausible range but none as to
the ultimate size of the world's population. Assuming no serious mortality
increases or widespread disruptions, projections vary from 8.5 billion to
13.5 billion or even higher. The lower bound assumes, unrealistically, that
the world as a whole will reach replacement fertility, approximately the two-
child family, within 20 years. The projection of 13.5 billion assumes
replacement fertility in 2040–2045.

Recent trends in urban growth, as we have already noted, suggest future
urban concentrations unprecedented in human experience. The United Nations recently published revised projections of urban population up to the
year 2000. The projection approach is "state-of-the-art," but the authors
note that the magnitudes projected go beyond our experience and may not
prove reasonable if human agglomerations of such size cannot be sustained.
Despite this appropriate caveat, the projections are instructive (see Table
9.3).

Table 9.3. Projected Population Increases in Major Cities, 1980 to 2000.

City/Region	Projected Pop. 2000 (Millions)	Estimated Pop. 1980 (Millions)	Change/ % Increase
Mexico City	31.0	15.0	+107%
São Paulo	25.8	13.5	+ 91%
Tokyo/Yokohama	24.2	20.0	+ 21%
New York/N.E. New Jersey	22.8	20.2	+ 13%
Shanghai	22.7	14.3	+ 59%
Beijing (Peking)	19.9	11.4	+ 75%
Rio de Janeiro	19.0	10.7	+ 78%
Greater Bombay	17.1	8.4	+104%
Calcutta	16.7	8.8	+ 90%
Jakarta	16.6	7.2	+131%

Source: United Nations Population Division, Urban, Rural, and City Population, 1950-2000, as Assessed in 1980, ESA/P/WP.66 (3 June 1980), p. 38.

The projections show five cities larger in 2000 than the largest human agglomeration ever experienced. Twenty-year increases of between 75 percent and 131 percent are projected for seven cities in developing countries, including Mexico City, São Paulo, Beijing, Rio de Janeiro, Greater Bombay, Calcutta and Jakarta. As the U.N. staff notes, some of the projected numbers, e.g., the 31.0 million for Mexico City, may not be attainable because of water supply problems, destruction of tree cover, transportation difficulties and other "natural or social limits to growth."[3]

Whether or not such magnitudes are attained, the growth of cities in many developing countries seems certain to be rapid, with consequent stresses on food and water supply, building materials, energy and so on. There are also likely to be repercussions for political organization and stability.

In the developed countries, with fertility already very low and the bulk of the population already using contraceptives, speculations about the future become more hazardous. U.S. fertility in the 1980s is expected to stay low by some experts[4] and by others[5] to rise dramatically. In the 1930s in

Europe, fertility as low as that in much of the developed world today produced exaggerated alarms about national decline that culminated in a profusion of pronatalist policies. Similar policies in much of Eastern Europe in the 1960s sometimes led to coercive childbearing, as the means of voluntary fertility control were denied. Whether such extremist responses occur, as for most political behaviors in the future, cannot be reliably predicted.

If fertility does stay low, and international migration is moderate, we can predict demographic effects with considerable accuracy. There will be a gradual increase in the average age of the population of such countries and a shifting of the "dependency burden" from nonproductive children of school age to nonproductive adults of postretirement age. As a result, more national resources will have to be allocated to the larger elderly group and fewer to the smaller young group.

Linking Population Trends to Other Problems

Population trends underlie all of the other problems under consideration here, although the links are not so direct or unmediated as sometimes claimed. The linking of population growth to food shortages has a long and controversial intellectual history, dating back at least to the essays of the Reverend Thomas Robert Malthus in the first half of the nineteenth century. In retrospect, we may conclude that Malthus's concepts were partially correct, but his predictions quite wrong. Technological and other improvements have allowed food production to more than keep up with the unprecedented growth of world population. Yet in principle, population cannot continue to grow indefinitely in a finite world, and signs of resource shortages and environmental stresses are already apparent.

Several theoretical efforts have been made to calculate the maximum human population sustainable by the world's agricultural resources, but such calculations often are highly stylized — even mechanistic. They consider arable land availability on a global basis, whereas in fact land is available only within sovereign nation states. They adopt idealistic assumptions of high average agricultural productivity, equal distribution of world food supply and a worldwide diet equal to that of Japan. They take into account no regional limits on water supplies or difficulties in moving surplus water from one region to another, no political or economic limitations on world commerce and no shortages of energy or fertilizer supplies. In short, these calculations contradict reality in fundamental ways. At the same time, their realism content is enhanced by their assumption of no dramatic quantum improvements in agricultural productivity. Were such completely unpredictable improvements to occur, they would counterbalance the unreality of the other assumptions.

The most recent effort to project population trends in relation to global resources is the *Global 2000 Report to the President,* produced by the Council on Environmental Quality and the Department of State. This three-year project sought to integrate a series of projections in various related sectors up to the year 2000, including climate, technology, food and agriculture, fisheries, forestry, water, energy, fuel minerals, nonfuel minerals and environment. The enterprise presented enormous technical and data problems, and its authors were forthright in admitting that they were only partially successful. In particular, they were unable in the time available to them to make the various projections fully interactive, so that changes in one sector could have full feedback effects in the other sectors. The report's authors conclude that the overall impact of these deficiencies is to "understate the severity of potential problems the world will face as it prepares to enter the 21st century."[6]

Despite these deficiencies, the report presents some interesting findings. It notes correctly that the momentum of population growth means that only moderate differences in population size by the year 2000 are possible, depending on the course of fertility in the coming two decades. The report makes the following projections: (1) Gross National Product (GNP) will grow more rapidly in developing countries than in developed countries. (2) Because of the lower starting point and the more rapid population growth in the developing countries, however, per-capita GNP increase in these countries will remain very modest in both absolute and relative terms. (3) Some developing countries, especially in Latin America (and presumably some OPEC—Organization of Petroleum Exporting Countries—nations), will improve significantly in per-capita GNP, although others will make few if any gains from present low levels. (Increases in India, Bangladesh and Pakistan, for example, are projected at 31 percent, 8 percent, and 3 percent, respectively, with all three countries remaining below $200 per capita in 1975 dollars.)[7] As a result of these trends, the report projects increasing per-capita income disparities between the wealthiest and the poorest nations.

With regard to food supply, the report summarizes its alternative projections with the cheering proposition that food production can continue to slightly exceed population growth up to the turn of the century, assuming no deterioration in climate or weather conditions. To achieve such growth, however, food production will require increasing inputs and technologies of a yield-enhancing, energy-intensive nature, such as fertilizer, pesticides, herbicides and irrigation. Such increased energy dependence of agricultural production has significant implications for the cost of food production, and the report projects a substantial increase in real food prices over the coming two decades, after decades of generally falling prices. Food production and consumption are projected to continue to be highly varied among the world's regions and nations, with rising food output of developing nations barely keeping ahead of rapid population growth. Furthermore, the high percentage of income already spent on food by hundreds of millions of people in poor countries suggests disturbing implications were the report's projected sharp increases in real food prices to take place.

Two points deserve comment here. First, in many developing countries, government policies seriously impede increased food production. In some countries, development policies provide direct or indirect subsidies to nonfood production but deny credit and other needed resources to food producers. Other countries use price controls on food products to keep urban consumer prices low and thereby stabilize the political structure, but such well-meaning controls may also produce losses for farmers and encourage rural-to-urban migration. Coupled with rapid population growth and unfavorable climatic trends, such policies have led to declines in per-capita food production in some developing countries over recent years, especially in Africa.

Second, it may be reasonably argued that global figures on food production are misleading, because the overwhelming bulk of food production is consumed locally. Efforts by the Food and Agriculture Organization to build a world food reserve have not yet succeeded, and even the large volume of international trade in foodstuffs consitutes only a small proportion of total food production. Hence, the primary goal of policy and technological innovation must be enhanced food production within the countries where demand is increasing rapidly, with international trade providing only marginal or emergency supplies.

Linking of Population Trends to International Security Issues

The AAAS Five Year Outlook Project mandates consideration of the relationships between population trends and international security issues. The literature on international relations contains a number of common hypotheses that oversimplify this complex issue:

- The larger a nation's population, the greater its actual or potential power.

- Population pressure on natural resources contributes to pressure for international aggression to obtain additional such resources.

- Nations with excessive population densities seek "living space" or "elbow room" via international aggression.[8]

Empirical analyses of international conflicts do not support most of these hypotheses. In general, population size and density appear to be underlying factors that may or may not contribute to international conflict. Much depends upon mediating political, social and economic factors, including stability of national political structures, distribution of available resources, technological and capital base of the nation, human capital available and patterns of consumption.

On the other hand, it is a truism that nothing can grow infinitely in a finite world. If population growth continues, it eventually will exceed the social, economic and political capacities of some nations. Hence, a summary assessment might be that eventual restraint on population growth is not a sufficient condition to assure internal and international stability, but that it is a necessary condition.

Problems of international conflict often are generated by internal instabilities within nations, as recent experiences in the Middle East demonstrate. Hence, it is also important to consider the effects of population change on internal stability as it relates to international relations. In this regard, several demographic trends demand attention. The first is the unprecedented rapidity of demographic change in many developing countries since World War II. A nation with a very substantial resource base may be able to support a much larger population, but if population size increases very dramatically, the rate of increase rather than the population size itself may contribute to instability.

A second component of high fertility is the distortions in the age composition that it engenders. High-fertility populations are also youthful populations, with typically 45 percent or more of their populations under the age of 15. Apart from the obvious problems such a concentration of young people presents for educational and other age-related services, such a steeply sloping age structure implies a very rapid growth in entrants into the labor force each year. The International Labour Organisation projections, for example, show increases of 600 to 700 million in the developing world's labor force in the next 20 years alone. To put these numbers into perspective, such an increase over two decades is larger than the entire 1980 labor force of the whole of the developed world.[9] In many developing countries already experiencing very high rates of unemployment and underemployment, such a rapid growth of young labor force entrants presents serious problems that can spill over into political instabilities. Such problems are further compounded in many of these countries by very rapid rates of rural-to-urban migration that contribute to even more rapid rates of labor force growth in urban areas.

Given the near certainty of rapid labor force growth in developing countries for the remainder of this century, labor-intensive development policies, especially in the rural areas, represent an important component of efforts to maintain national coherence and internal stability. To the extent instability and dissolution in such countries spill over into the international sphere, as happened recently in Iran, policies favoring intensive and broadly based job generation also favor international security interests. Although developing countries themselves must make any decisions favoring such policies, developed countries, such as the United States, can make these policies more attractive and feasible, through trade and tariff policies that favor imports produced in labor-intensive industries. International political support can also encourage governments to move toward such domestic policies.

Implications for U.S. Science and Technology

Science and technology have already contributed impressively to recent population trends and can be expected to continue to do so. The rapid acceleration of population growth in developing countries after World War II owed much to improvements in health, nutrition and sanitation, due in some (perhaps, great) measure to science and technology. The sustainability of the so-called population explosion (which, in fact, bears more resemblance to a speedy glacier than to a bomb) owes much to the improved productivity of agriculture and technological innovation, as well as the capacity to convert abundant energy resources themselves (also a product of technological innovation) into edible calories.[10] Equally important were improvements in technologies of significance to public health, ranging from sanitary water systems, to biologicals such as vaccines, to improvements in internal and international communication and transportation that diminished the deadly impact of localized food shortages.

Improvements in communication and transportation have also contributed to internal and international migration. Isolated rural populations discovered the relative attractiveness of life in urban areas or in other countries by listening to transistorized radios and watching television programs brought to them by communication satellite. At the same time, the improvement of internal road, rail and air networks has facilitated movements to the urban areas, and the increased availability and declining real price of international air travel following the development of modern aircraft technologies have sharply reduced the nonlegal barriers to international migration. Finally, the availability of satellite communication has brought vividly to the attention of the world the plight of millions of miserable refugees starving and dying on the high seas or in temporary encampments. In the way that television is said to have affected perceptions of the war in Vietnam, so, too, has it changed public images of refugee problems.

Science and technology have also contributed greatly to fertility declines. While some form of fertility control has been available in most human societies, it is often forgotten that highly effective contraception is a development of only the past 20 years; the first oral contraceptives were not widely marketed until the early 1960s, and the intrauterine device (IUD) was not widely available until the same decade. Similarly, safe and acceptable male sterilization via vasectomy (the most popular fertility control method in some countries) did not become common until the 1970s, although tubal ligation for females was in use earlier. There have also been substantial reductions in the health risks of induced abortion due to technological advances.

Finally, science and technology have also contributed greatly to our collective understanding of population change and its impacts. Demography and some of the social and statistical sciences have, over the past 30 years, provided new and powerful tools by which we are now able to detect and estimate demographic rates—in some respects analogous to the technologies that have improved our capacities to assess agricultural potential, measure air and water quality and even predict the weather. Such demographic tools now allow indirect estimation of demographic rates among populations whose births and deaths are not registered and in nations that have never conducted an adequate census. Other important scientific advances have contributed to our understanding of the factors affecting age composition and the momentum of population growth, the patterns of marriage behavior and the relationships of mortality change to fertility behavior.

The Outlook

As to future contributions, the capacity of world agriculture to accommodate to projected 40 percent increases in population in 20 years will depend heavily upon the contributions of both U.S. agricultural production and scientific expertise. Similar contributions can be made on the mortality side, via intensive work on tropical diseases that continue to be large-scale killers and maimers, such as river blindness, schistosomiasis, diarrheal diseases and cholera.

With regard to fertility, it is evident that the array of contraceptive methods presently available, although a substantial improvement over those before 1960, is inadequate to the needs of large numbers of people and nations. As has often been pointed out, oral contraceptives are relatively nonspecific in their modes of action and have side effects that make them inappropriate to the needs of many people desiring effective fertility control. IUDs equally have notable limitations (in fact, their mode of operation is but little understood), and available sterilization methods are less acceptable than they would otherwise be because they are substantially irreversible.

Furthermore, the diversity of social, economic, cultural and religious settings in the world today means that a method that is highly preferred in one setting may be unacceptable in another. Even within societies, individuals require different contraceptive techniques; indeed, the same individual may require a variety of methods through his or her lifetime – a useful illustration of the diversity of contraceptive demands. If we define "marital" to include stable consensual unions, we can describe four stages of the individual's reproductive life cycle:[11]

1. premarital;
2. delay (postmarital, pre-first birth);
3. spacing (post-first birth, before completion of fertility);
4. completion of wanted fertility.

The contraceptive characteristics most suitable for each of these stages are presented in Table 9.4. Such variety of individual needs, coupled with the diversity of national, religious and cultural settings, suggests that there can be no such thing as "the ideal contraceptive"; what is required is an array of methods with differing attributes, collectively providing an adequate scientific and technological response to the needs presented by human diversity.

Contraceptive technology advanced greatly in the 1960s, but little since, and there are few promising methods on the immediate horizon. The scientific and technological pipeline is a particularly long one in the field of fertility control, given the appropriate concern of governmental regulators as to the safety of methods that may be used by millions of healthy young adults. Over the next 5 to 10 years, only a few potential improvements are in prospect – a subdermal implant for slow release of contraceptive hormones may prove effective and safe, and some improvements may be made to existing IUD technology. Although science and technology often confound the most reasonable predictions, there are, at present, no great anticipations of new methods to fill some of the obvious gaps: effective male contraceptives other than condoms and reversible methods of voluntary sterilization. The problems are not in the realm of technology or product development but, rather, result from our very limited understanding of the remarkably complex process of human reproduction. Yet in the recent past, scientific attention to this area has been modest; the study of human reproduction, prominent in the 1930s, nearly died out in the 1940s and had to be resurrected in the 1960s. As a result, it is a Johnny-come-lately that remains a minor claimant on government research resources.

Table 9.4. Characteristics of Contraceptives Related to Life Cycle Stages.

Stage	Characteristics
1. Premarital	Relatively irregular and infrequent exposure. Intercourse-related methods (particularly postcoital) somewhat more acceptable than in delay and spacing stages. Serious consequences for contraceptive failure. Limited knowledge of and access to fertility control. Limited independent access to medical system, hence nonmedical delivery is preferable. Reversibility highly important.
2. Delay: Postmarital, Pre-first birth	Frequent exposure. Relatively moderate consequences for contraceptive failure. Relatively short period of protection required. Methods where application is independent of intercourse are highly desirable. High acceptability and convenience important. Method delivery via medical system is less undesirable than in premarital stage due to readier access to medical system. Reversibility highly important.
3. Spacing: Post-first birth, Precompletion	Frequent exposure. Moderate consequences for contraceptive failure. Long time span (as sum of separate birth intervals) of protection required. Reversibility somewhat less important than in delay stage.
4. Completion of wanted fertility	Long time span of protection required. Less frequent exposure than in delay and spacing stages. Serious consequences for contraceptive failure. Intercourse-related methods somewhat more acceptable than in delay and spacing stages. Acceptability and convenience less important than in earlier three stages. Reversibility less important than in three previous stages.

Note: Characteristics discussed here are average characteristics and need not apply to any particular individual in any stage.

Source: Roy O. Greep, et al., Reproduction and Human Welfare (Cambridge, Mass. and London: MIT Press, 1976), p. 71.

In spite of vigorous rhetoric to the contrary, available evidence shows that the overwhelming majority of Third World people live in countries whose governments openly declare their desire to lower rapid rates of population increase. In the most recent survey by the United Nations (in 1978), such countries comprised fully 82 percent of the population of the developing world, including most of the largest (for example, China, India, Indonesia, Bangladesh and Pakistan). A relatively large number of nations with small populations, especially in Latin America and sub-Saharan Africa, however, reject the need for such a demographically oriented policy. Hence, the one-nation, one-vote structure of the United Nations and other international forums sometimes conveys a less prominent commitment to reducing rapid population increase in the developing nations than is, in fact, the case.

The ability of developing countries to lower their population growth rates as a matter of public policy depends heavily upon improved knowledge of the social, economic and cultural factors favoring fertility decline and upon improved skill in implementing service programs that must reach literally millions of couples. The causal mechanisms underlying past fertility declines are imperfectly understood, even for the developed countries. Existing governmental strategies aimed at encouraging fertility decline range widely:

- development policies aimed at enhancing presumed indirect factors favoring fertility decline (Egypt);

- policies for directly providing knowledge and means of fertility regulation (India, China, Mexico, Bangladesh and Indonesia, to name a few);

- the use of economic and other incentives affecting individual fertility behavior (Singapore, China);

- official support for direct application of "pressure " or "persuasion" (China).

If future policies are to be more effective, there is much to be learned about the impacts of such an array of strategies in diverse settings, and the tools of social science and evaluation research are the only means for such learning. Such efforts can be highly cost-effective, as a modest research investment can result in substantial improvement in the implementation of expensive large-scale programs and can suggest new or additional strategies that may prove more effective in a given social, economic or cultural setting.

It is commonly believed, and often pronounced, for example, that declines in infant and child mortality are both necessary and sufficient to lower fertility; hence, in many settings, population policies concentrate heavily on maternal and child health services. Scientific evidence on this question, however, is mixed. Historical analyses of the European fertility transition suggest that mortality declines were not consistently important explanatory factors.[12] Evidence from developing countries suggests that fertility response to infant and child death is only partial[13] and may be larger in some settings than in others.

Equally unknown is the nature of factors that have led to dramatic increases in marriage age that in some countries have accounted for a large percentage of birth rate declines. Enhanced understanding here could provide new and effective policy levers for government officials.

Finally, knowledge of the pattern, magnitudes and causes of internal and international migration is notoriously deficient; as these population movements grow in size and impact, it is evident that coping with them will require the illumination that comes only with scientific analysis.

U.S. scientific and technological innovation ranks high as both initiator and moderator of recent population problems. Many of the effects of rapid population growth are only now coming to be felt, as the large surviving generations born in the 1960s and 1970s reach adulthood and seek employment and lives of dignity. It seems certain that meeting these human needs and moderating present rapid rates of population growth will require enhancement of U.S. scientific and technological contributions over the coming decades.

Notes

1. For an authoritative yet concise assessment, see W. Parker Mauldin, "Population Trends and Prospects," *Science,* vol. 209 (4 July 1980), pp. 148–57.

2. See, for example, the controversy and reportage that followed publication of Amy Ong Tsui and Donald J. Bogue, "Declining World Fertility Trends, Causes, Implications," *Population Bulletin,* vol. 33 (September 1978).

3. United Nations Population Division, *Urban, Rural, and City Population, 1950–2000, as Assessed in 1980,* ESA/P/WP. 66 (New York: U.N. Department of International Economic and Social Affairs, 1980), p. 57.

4. Charles F. Westoff, "Marriage and Fertility in the Developed Countries," *Scientific American,* vol. 239 (December 1978), pp. 51–7.

5. Richard A. Easterlin, *Birth and Fortune* (New York: Basic Books, 1980).

6. Council on Environmental Quality and Department of State, *The Global 2000 Report to the President: Entering the Twenty-first Century* (Washington, D.C.: Government Printing Office, 1980), vol. 1, p. 7.

7. Ibid., p. 13.

8. See U.S. House of Representatives, Select Committee on Population, *Population and Development Assistance,* Ninety-fifth Congress, Second Session (December 1978), pp. 20-22.

9. International Labour Organisation, *Labour Force Estimates and Projections, 1950-2000,* 2nd ed., vol. 5 (Geneva, 1977), Table 6.

10. As an indication of how rapidly perceptions can change, it is worth recalling that only a decade or so ago there were popular and scholarly discussions of developing new technologies by which cheap petroleum from the Middle East could be converted directly into edible proteins during tanker shipment.

11. Roy O. Greep et al., *Reproduction and Human Welfare* (Cambridge, Mass., and London: M.I.T. Press, 1976), pp. 68-73.

12. Ansley J. Coale, "The Demographic Transition Reconsidered," *Proceedings of the International Union for the Scientific Study of Population,* vol. 1 (Liege: International Union for the Scientific Study of Population, 1973), pp. 53-72; and Michael S. Teitelbaum, "Relevance of Demographic Transition Theory for Developing Countries," *Science,* vol. 188 (2 May 1975), pp. 420-25.

13. See, for example, Samuel H. Preston, ed., *Effect of Infant and Child Mortality on Fertility* (New York: Academic Press, 1977).

10
International Security
Implications of Materials and
Energy Resource Depletion

William A. Vogely

Materials and energy depletion have been a continuing fear of mankind since the industrial revolution. Throughout the nineteenth and twentieth centuries, thoughtful observers have warned of the ultimate exhaustion of the materials and energy resources upon which society is based. Classical economists predicted a steady state of no growth and labor at subsistence wages. In 1866 Stanley Jevons wrote in the preface to the second edition of *The Coal Question:*

> Renewed reflection has convinced me that my main position is only too strong and true. It is simply that we cannot long progress as we are now doing—not only must we meet some limit within our own country, but we must witness the coal produce of other countries approximating it to our own and ultimately passing it . . . our motion must be reduced to rest, and it is to this change my attention is directed.

Jevons's words, in a different context, are echoed in 1972 in the introduction to *The Limits to Growth:*

> If the present growth trends . . . continue unchanged, the limits to growth on this planet will be reached sometime within the next 100 years. . . . it is possible to alter these growth trends and to establish a condition of ecological and economic stability that is sustainable far into the future.[1]

The subject of materials depletion is an extremely broad and multifaceted one. This paper limits itself to looking at the implications of materials and

William A. Vogely is professor of mineral economics, Pennsylvania State University, University Park, Pa.

233

resource depletion for international security. Thus, it will ignore many interesting and important areas, such as conservation and environmental issues concerning the production and use of materials and energy. It will, however, look at the process of resource depletion in order to state clearly the nature of the depletion problem.

The Process of Resource Depletion

Resource concepts are semantically difficult because the terminology used to describe resources is confusing and the same words mean different things to different people. The literature is full of resource life indexes that divide resource stock by either annual or cumulative production based upon an annual rate of growth and measure the number of years remaining to each resource before it is exhausted. These indexes misunderstand, perhaps deliberately, the nature of resource supply. Whether presented with sophistication and understanding or presented in ignorance, the resource life indexes represent a fundamental misstatement of the problem of depletion. Resources flow into the economy — they are not an inventory to be used over time.

Depletion of a natural resource occurs at three distinct levels: (1) single deposits; (2) the replacement of deposits in the production function; and (3) at the ultimate occurrence of the resource in the earth. Much of the misunderstanding about resource terminology has occurred because words derived from one of these levels are applied to another level. [2]

Depletion of a Resource Deposit

Natural resources occur in nature in deposits that have unique chemical, physical and locational characteristics. Some deposits of mineral resources are economic to produce in relation to the markets for their product. These deposits may be developed into producing sites: mines, if the resources are solid; or fields, reservoirs or wells, if the resources are liquid and gas. A known deposit that is capable of being produced today is called a reserve. These reserves will be produced through time from the deposit. The deposit may be extended through exploration; the reserves, through additional capital investment. Material produced from that deposit, however, will not be replaced in the deposit; thus the deposit will begin to be depleted as it is produced. Depletion of a deposit simply means that, for every ton produced, there is one ton less left to produce. As a single deposit is used up, the cost of production from that deposit tends to increase. The deposit is considered "depleted" when it is no longer economically attractive to continue production. The deposit will then be abandoned and, some would say, it is exhausted.

It is important to note, however, that virtually no resources have been physically exhausted. For a typical oil field, an average of over 60 percent of the original oil remains in the deposit upon abandonment. In the case of the nonfuel resources, mine sites are abandoned because the remaining ore does not justify further investment to develop it. But if the investment again becomes worthwhile, the oil field or mine site may be reopened. With high prices of gold, for example, hundreds of abandoned mine sites in the West are being opened; and, with the increased price of oil, abandoned wells are also being produced.

Reserves are determined and measured in terms of a specific deposit. Deposits are abandoned for economic reasons and not because their contents are literally exhausted or reduced to zero.

Replacement of Deposits

Except in geologic time, the distribution of energy and materials in the earth's crust can be taken as fixed. In this distribution, deposits come in all sizes, shapes, grades and chemical characteristics. These deposits are discovered through exploration. The deposits that are profitable to develop become producing mines and contain reserves. As it ceases to be economically attractive to extract resources from a deposit, that deposit is replaced by a new one. New deposits are discovered by investment in exploration and become producing mines through further investment. The replacement of deposits is a function of exploration and of investment to develop the deposit. Deposits frequently remain undeveloped because the economic cost of developing them is not attractive, given the markets for the commodities. Thus, deposits are replaced either when a new, economically attractive deposit is discovered or when technologies for developing known deposits at an attractive cost are developed.

At this second level of consideration, depletion can be said to be occurring when the replacement deposits are of higher real cost per unit than the depleted deposits that they replace. This is the aspect of depletion that has been discussed most thoroughly in the literature. In their path-breaking book, *Scarcity and Growth*, H. J. Barnett and C. Morse tested the process of depletion of replacement deposits by positing that if it were occurring, the real costs or real price of materials should be rising through time.[3] They were not able to prove this hypothesis and, in fact, found that such real costs were declining in the period of 1870 to the 1950s. Recent work by V. Kerry Smith and others has weakened that conclusion with respect to the period following the 1950s.[4] It is depletion in this sense, however, that underlies most of the literature with respect to resource exhaustion.

Depletion of the Resource Base

All of the elements in the upper earth's crust, water and atmosphere are considered the resource base. It is theoretically impossible to deplete these resources. Mankind is only able to redistribute, not destroy, them. In the case of the nonfuel resources, production concentrates them from their natural occurrence and, in a sense, creates new mines from which they can be reclaimed through recycling. In the case of the energy resources, use does reduce the energy potential contained in those resources and, in that sense, increases the entropy within the universe. Clearly the forces of geologic processes are to level the earth, and in time the energy flow will reach an equilibrium state of zero. The time spans for such events, however, are well beyond the projected and possible survival of mankind.

From a global point of view, resource deposits in nature can be ranked by the cost of producing them under any given state of technology. Such a ranking, although impossible to quantify, would present a picture of a stepwise increase in cost as resources with different economic dimensions are used. At one end of the spectrum would be the resource content of sea water or of common rocks, the supply of which is "inexhaustible." The cost of obtaining any given mineral element from these ultimate resources may be infinitely high; but these resources are, nevertheless, physically inexhaustible.

Summary

As Zimmerman has pointed out, "resources are not, they become."[5] The principles already sketched underlie the current orthodox classification of resources along the double axis of economic availability and geologic identification. The current resource classification system used by the federal government, presented in Figure 10.1, illustrates these concepts. This basic idea of resource categorization has many variants, and, of course, there is much discussion concerning what kind of numbers to put in the various boxes. The process of resource depletion is both an economic and a geologic phenomenon. It is economic in the sense that any deposit will be abandoned when continued production is no longer economically justified; the replacement of that depleted deposit depends both on geologic occurrence of deposits and the economics of additional capacity; and, finally, the limit on further production is always an economic, not geologic, phenomenon.

The Concept of Resource Adequacy

The concerns expressed by the authors quoted in the introduction relate not to exhaustion as a phenomenon but to the fact that a decline in resource availability to mankind will impose real limits to the quality of life of mankind. This concern, which broadens the scope of the analysis from the economics and geology of resource deposits, raises the problem of resource adequacy. By definition, adequacy must be measured in terms of objectives. Thus, the subject of resource adequacy has both a supply and a use side.

Figure 10.1. Classification of Mineral Resources.

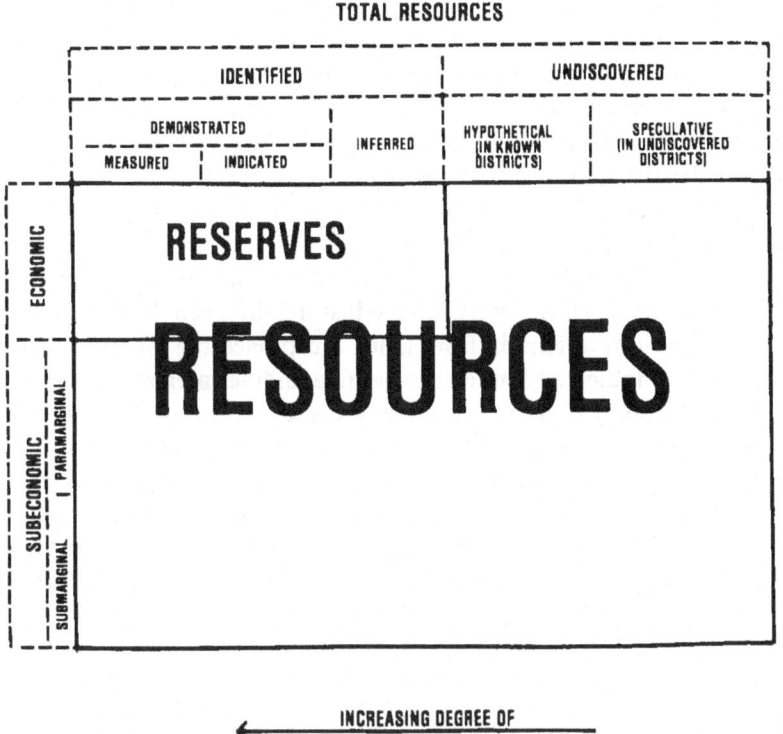

Several general measures of resource adequacy have been proposed. T. Page, for example, suggests that we use constant costs of resource availability through time as a test of resource adequacy. V. K. Smith tries to develop a scarcity index of resource adequacy.[6] Others define it more narrowly, in terms of resource adequacy for a three-year war, as defined by government policy with respect to strategic stockpiles. Still others look at resource adequacy from the point of view of whether a given resource is adequate to allow society to undertake actions to replace its use with another, which is the current underpinning of energy policy with respect to liquid fuels. All of these concepts have a common analytic structure. They involve adequacy as measured by supply with respect to an objective or demand for the resource. Adequacy always has a supply and a demand side.

Supply Side of Adequacy

Virtually all analyses of the supply side of adequacy start with some measurement of the size of the various resources categories shown in Figure

10.1. As we have already noted, transformation of resources to reserves is the fundamental issue on the supply side of resource adequacy. Recognizing this, we see at once that the critical variable is the quantity of reserves that can be ultimately developed, given the economic costs. But this quantity is unknown and, at the current state of our knowledge, unknowable. All attempts to estimate what R. G. Ridker and W. D. Watson call "prospective reserves" involve the application of the current state of human knowledge to predict or project unknown quantities.[7] There are three primary methodologies being used to make such estimates.

The most familiar method relates the remaining volumes of prospective reserves to the rate at which reserves have historically been developed and used. These time rate methods flow from the pioneering work of King Hubbert, and they indicate a limited prospective reserve category for oil and gas, uranium and some other major mineral commodities.[8] A second technique uses a geologic or geographic analogy, whereby the material and fuel content of a known geologic environment is assumed to be replicated in all such geologic environments in the earth's crust, or, in a more general sense, the material and energy content of a given geographic area is assumed to be replicated in other equal-in-size areas. The third method is to ask the experts and develop a probability range around an estimate.[9]

The usefulness and accuracy of each of these methods are, of course, open to sharp attack, on the grounds that we cannot estimate a phenomenon when our basic scientific understanding of that phenomenon is flawed. At the beginning of the energy crisis, when it became very important to develop an understanding of the future availability of petroleum and natural gas in the United States for public policy purposes, the Federal Energy Administration asked a group of distinguished statisticians to look at the alternative methods for estimating the ultimate reserves or producibility of oil and gas in the United States. These statisticians, working independently, each arrived at the same conclusion that none of these estimating techniques was statistically reliable.[10] The Geological Survey has developed a model for the availability of petroleum that shows the absolute necessity of starting with a scientifically justified model of the occurrence of deposits by size distribution and other characteristics in the earth's crust. This has to be followed with knowledge of how many of these deposits can be discovered and at what cost through exploration and so on through the development and production stage.[11]

The import of the preceding paragraph is that *we must give up any hope of developing a scientific model of future availability at what cost for any material or energy resource.* At best we can take the first element of resource classification, reserves, as the minimum that will become available, by definition, at current real price. Beyond that, we can—with decreased certainty—estimate geologic discoveries and technological advances. It must be understood, however, that any figure or range of figures so developed misstates the fundamental concept of the supply of resources. The supply of resources is, in fact, a flow of resources to the economy. It is not the exhaustion of a fixed stock. Whether or not resource depletion in the sense of rising costs occurs is a function of future technology in exploration and in production.

Knowledge of future supply side availability is very limited. The origin and real costs of supplies for the next decade are now known with very small margin of error. The margin of error consists primarily of the political availability of known resources, that is, cutoffs in supplies arising from political constraints, such as war or embargo. In addition, there is the unknown probability of major natural disasters. Finally, there is a small probability at the margin that major new deposits or technical advance could change the supply situation within the next decade. The latter is extremely unlikely, given the long lead times necessary to develop the productive capacity and infrastructure involved in major new material and energy projects. For many commodities, current reserves contain quantities that still will not be used before the end of the century. For others, current reserves will not last this decade if demand for them continues at current levels.

Demand Side of Adequacy

The determinants of resource demand can be categorized into seven major variables:[12]

1. Demographic variables, such as size, rate of growth and age and sex distribution of population; number of households; and labor-force participation rates.

2. Standard of living, usually represented by per-capita gross product.

3. Style of living, such as the pattern of preferences in consumer goods and transportation services.

4. Geographic distribution of population between urban and rural.

5. Technological structure, that is, the means by which goods are produced from resources.

6. International trade relationships.

7. Institutions and policies, for example, environmental requirements.

The above factors affect the demand for total resources. The use of any single resource is the result of demands for final goods, the technology of production of each good and relative prices.

It is clear that many of these variables are useless for projecting material demands, because their projection is difficult, and the relationship between them and specific resource demands is very complex and uncertain. Most analyses of projected resource demands rest upon some simple assumptions of the relationship among specific resource demand, levels of production and, in some cases, the price of the resource. The important point to be made here is that, just like the supply side of adequacy, the demand side of adequacy is essentially unknown and unknowable, as it depends irrevocably upon the development of future technologies and price relationships.

Recent Studies of Resource Adequacy

There have been three major studies, published in 1979 and 1980, that have attempted to measure in quantitative terms the adequacy of materials and fuels resources.[13] These studies carefully examine the evidence, make what the authors consider to be conservative projections and draw general conclusions. All three studies indicate that depletion, as measured by its economic dimension of increasing costs, does not present a challenge to resource adequacy for a minimum of three decades. The studies do conclude, however, that, in the area of energy, society faces a transition from its current sources to alternative sources, and that the effect of resource depletion on the quality of life rests upon the successful conduct of that transition.

The concept of resource adequacy rests upon the conjunction of materials and energy supply with materials and energy demand. The essential problem is whether quantities will be available at any given price level to meet requirements at that price level. The future characteristics of economic availability are unknown and, at the current state of knowledge, unknowable. Therefore, any projections of availability as a function of price through time for mineral resources are highly uncertain. The same can be said for the projection for the use of mineral resources, that is, the demand side. Both the supply function and the demand function are subject to extremely complex determination, and prediction of the factors determining each through the future is virtually impossible. The predictions progressively lose credibility as a function of future time. The situation is not, however, as bleak as it seems. If the adequacy of resource availability is seen as a process rather than a point estimation, it is possible to develop strategies addressed to the process itself which have implications and viabilities beyond our knowledge of future outcomes.

Worldwide Distribution of Reserves

Reserves are developed in response to economic incentives, that is, the prospects of returns from development of mineral resources. The factors that determine which deposits will be discovered and then developed into productive reserves involve calculations not only of the costs of developing the deposit itself but also of transportation and marketing of the output. The closer a given deposit is to the marketplace, the greater the likelihood it will be developed. Deposits in remote areas must carry a substantial premium in the sense of economic rents to justify the transportation and other costs involved in their development.

It is not surprising, then, that mineral developments have been located near industrial markets. The importance of location to production is perhaps best illustrated by the steel industry, where location has been the result of the confluence of the basic raw materials, energy and markets. In the United States, the original centers of production were in the Pittsburgh area. These centers used river transportation for the coal, iron ore, limestone and other inputs to the process and served the emerging industrial complex of Pennsylvania and Ohio. As the iron ore supply shifted northward to Minnesota, a second complex, centered in Gary, Indiana, was generated along the shores of Lake Michigan. The Japanese steel industry takes full advantage of low-cost ocean transportation for all of its raw materials and much of its product. The total cost of supplying a market is critical in decisions about where to locate production.

Of course, the location decision is affected by the geology of the mineral resource itself. Given all other factors, however, exploration will tend to be concentrated in those areas where development would be relatively easy. An oil reservoir might be a bonanza in Oklahoma, for example; yet the same oil reservoir in offshore Nova Scotia might not be a commercial find.

Resources can be categorized with respect to the importance of the market and transportation systems in their location. For the construction materials that make up in total bulk most of the materials society uses, development is almost entirely market oriented. At the other extreme, the ferroalloy metals, which are geographically scarce and measured in pounds rather than in tons, are developed where they are found. Most major resources lie between these extremes.

Commodities produced far from their markets create international security implications. The major one, of course, is petroleum, but cobalt, chromium, platinum group metals, manganese, copper and bauxite are also important. Each of these is briefly discussed below.

Petroleum. The United States and other Western industrialized countries use petroleum primarily for transportation but also for industrial processing and household energy. The bulk of petroleum entering into world trade is subject to the actions of a cartel that has been successful in raising the price of petroleum in the world markets. The success of the cartel flows from two fundamental characteristics of the petroleum market:

- Because the demand for liquid petroleum is technologically fixed in its transportation uses, at least for a significant period of time, rapid substitution away from liquid petroleum in transportation is virtually impossible.

* The search for new petroleum reserves is expensive and carries a long lead time, and so the development of reserves outside of the cartel's control is a relatively slow process.

These two factors combined have permitted the cartel to raise the price of petroleum on world markets by an order of magnitude over the past 10 years. The Arab portion of the Organization of Petroleum Exporting Countries (OPEC) cartel did use the "oil weapon" and imposed an embargo in 1973, and supply from the Middle East has been interrupted periodically by wars since 1950. Thus, petroleum is an example of a concentration of world reserves under the control of a cartel and subject to supply interruption by deliberate action or as a result of political developments.

Cobalt. Over 40 percent of the world mine production of cobalt comes from Zaire, which has well over a third of the world's reserve base. Cobalt has a variety of uses, but its most important one, from the point of view of international security, is for turbine engines in aircraft. It is produced as a by-product of copper, and the price is set by the Zairian source. Total world production is only about 35,000 tons.

Chromium. Chromium is an essential ingredient for the making of stainless steel. Thirty-five percent of world production comes from the Republic of South Africa, which also has two-thirds of the world reserve base.

Platinum. Half of the world production and three-quarters of the world's reserve base for platinum group metals is in the Republic of South Africa. Virtually all of the remainder is in the Soviet Union. A major use of platinum that raises international security implications is its use as a catalyst in the refining of petroleum. It is also used for emission control in automobiles in the United States.

Manganese. Manganese is an essential ingredient, under current technology, for the making of steel. The Republic of South Africa supplies a fifth of the world's mine production but over 40 percent of the free world production. South Africa contains three-quarters of the free world reserves and about a third of the world reserves of manganese.

Copper. Copper reserves are much more broadly distributed than the other commodities listed above, but copper does not enter into world trade in significant volumes. The largest producers are the United States, Chile, the Soviet Union, Canada and Zambia, in decreasing order. On the reserve base side, the reserves are held approximately one-fifth by Chile, another fifth by the United States, followed by Russia, Zambia and Canada, each of whom has less than 10 percent.

Bauxite. Bauxite is the ore for aluminum. Bauxite ores are widely distributed throughout the world, but bauxite is a commodity in which the geographical separation between the ore producers and the metal producers is pronounced and virtually all of the bauxite enters into foreign trade. The largest producer is Australia, which accounts for about 30 percent of the world's production. Guinea and Jamaica each produce about 15 percent, and individual countries drop off sharply from that level. On the reserve side, Guinea has approximately 30 percent and Australia 20 percent, followed by Brazil and Jamaica with about 10 percent each.

Summary. Distribution of reserves and productive capacity within the world arises from geologic and economic factors. As indicated above, geology has played the most important role in production of petroleum and some of the ferroalloy metals. For most other materials the primary factor has been economics, not geology.

Resources and Economic Development

Abundant natural resources have played a major role in the economic development of the nations of the world. At the time of the industrial revolution, the confluence of energy and material availability was a determining factor in the location of major industrial activities. Clearly, the emergence of Great Britain, Western Europe, the United States and Japan as major industrial powers is based upon a natural endowment of energy and material resources or access to ocean transportation to permit their acquisition relatively cheaply.

The developing countries now look upon resources as a major means of facilitating their economic development. The export earnings flowing to the oil producers, greatly enlarged by their cartel action, have provided a clear example of the transfer of wealth from the industrialized countries to the raw material producers. In addition to petroleum, copper has played a major role in Chile and Zambia and is looked upon as a major contributor in such countries as Papua New Guinea and Panama. The Republic of South Africa, which is blessed geologically with a disproportionate endowment of manganese, chrome and platinum group metals, has used these materials plus gold and diamonds as a major source of its wealth.

Among the industrialized countries, the Soviet Union is least dependent upon the international flow of goods for its mineral and energy supplies. In part, this is due to resource endowment, but it is also due to deliberate government policies. The Soviet Union, for example, does not rely on imports of bauxite for its aluminum and thereby imposes substantial additional costs for the production of aluminum metals. At the other extreme, Japan has virtually no natural resources and is, therefore, almost entirely dependent upon the rest of the world for imports of materials and energy for its industrial production. Between these extremes, the United States lies closer to Russia, and Europe lies closer to Japan.

The nonindustrialized areas of the world depend upon raw material exports as their major earner of claims to goods and services, and the industrialized countries depend upon raw material imports to maintain their economies. This fact creates, in essence, a bilateral monopoly bargaining position between the raw material exporters and the industrialized countries of the West. The exporters have a strong bargaining chip in that the industrialized societies, certainly within short time spans, cannot operate without the materials and energy they produce. On the other hand, unless these materials and energy are sold to the industrialized countries, the exporters will not be able to enjoy the returns from them and will suffer dramatically in terms of wealth.

Role of Depletion

The time frame of this paper is 5 years or, generously, the decade of the 1980s. Depletion of materials and energy resources during the next 10 years will not significantly affect the flows of trade and the international security aspects of materials and energy availability to the United States. In this sense, depletion is simply unimportant within the context of this analysis.

In two of the specific commodities we have already discussed, however, historical depletion is important to the current situation. The United States was the first large developer and user of petroleum in the world and has for many years maintained a position as either the major or a major producer of copper. On a relative basis then, the geologic deposits of petroleum and copper available within the continental United States have been depleted relative to deposits occurring in other portions of the world. To the extent that intensive exploration for these resources has discovered relatively high-grade and easily found resources within the land available for exploitation, future discoveries will be relatively less probable in the United States than in the rest of the world. The same situation applies to a wide range of other materials found in the United States, such as zinc, potash and sulfur.

For the other materials discussed above, however, geologic factors have prevented the United States from ever enjoying comparative advantage in their production. These materials are relatively scarce ones (except bauxite), and the best deposits simply do not occur within the boundaries of this country. Thus, the development of these materials has occurred outside of the United States, and depletion, as such, has played no role in that development.

From a world perspective, issues of resource depletion may have implications centuries from now. Much fundamental work is now being under-taken to try to understand the nature of substitution between resources, capital and labor in the production function of society.[14] If, in fact, this substitution is limited, then the question of resource constraints' placing a major limit on the growth and welfare of society is perhaps still open. Depletion in this sense is of great theoretical interest but of little practical interest in terms of resource availability over the next 10 years.

International Security Issues

Two major issues arise from the geographic and economic distribution of materials and energy raw materials in the world:

1. How can the industrialized world deal with supply interruptions?

2. Does relative depletion of resources in the industrialized countries jeopardize the comparative advantage of the manufacturing sectors of these economies, leading to short-term transition problems and long-term deterioration in their terms of trade?

The United States has been addressing the first issue in several contexts almost continuously since World War II. A large number of presidential commissions and special studies have dealt with the so-called critical materials problem. At present, for example, serious discussions concerning the "resource war" with southern Africa label the Republic of South Africa, in particular, as the "Saudi Arabia" of materials.[15] In the United States, the president's draft report on nonfuel mineral policy identifies the concentration of productive capacity and reserves in southern Africa of chromium, cobalt, manganese and the platinum group metals as a major issue with respect to short-term interruption of supplies.[16] The so-called energy crisis is precisely of the same nature, in that interruptions of petroleum have immediate and serious consequences to the industrialized countries.

A second consideration, separate but related to supply interruptions, is the economic terms upon which these internationally traded materials become available to industrialized societies. Oil is the significant material that raises this issue, simply because of its importance in the world economy. Cartels have been tried in bauxite, and there is producer pricing of both cobalt and chrome. These materials have economic values in the small range of millions of dollars, however, rather than in the tens of billions, and thus the impact of price rises on the overall economy is relatively trivial.

The issue of loss of comparative advantage is a serious one for the future of the United States, in particular. The U.S. industrial base was built while the United States had access to cheap and conveniently placed natural resources. Several basic industries are showing signs that they have lost their international comparative advantage and, perhaps, their absolute advantage. The problems of the steel industry, the automobile industry and the textile industry are symptomatic of this development. So long as the decline occurs in an orderly way and does not generate substantial local income distribution problems, its impact may not raise serious issues. But in the case of steel and automobiles, in particular, the transfer of labor and capital from a declining basic industry to a growing sector of the economy generates very substantial economic problems and perhaps international security implications.

The two problems are related and aggravated, in part, by relative resource depletion. Such relative depletion contributes to the loss of comparative advantage, which in turn contributes to increasing import flows of basic commodities, such as steel.

Alternative Strategies

The issues of supply interruption and loss of comparative international advantage are interrelated in a direct but subtle way. If society decides to solve the first problem in ways that substantially increase the cost of raw material supplies, it will exacerbate the second problem of comparative advantage. Energy provides a clear example. The industries suffering a loss of comparative advantage, such as the primary metals and basic manufacturing industries, have energy as a major cost of production. Thus, if the United States decides to solve its energy supply problem by imposing substantial costs on U.S. consumers above those borne by the other industrialized countries, this action in and of itself will accelerate the problems arising from a declining industrial base. This implies that the strategies for attacking these two problems must be considered together. Any viable solution must take full account of their interdependence.

The Self-Sufficiency Strategy

This is presented in the hopes that it will be perceived as a straw-man argument. It must be taken somewhat seriously, however, because the initial response of the federal government to the Arab oil embargo of 1973 was to proclaim a drive for "energy independence." It also gains credence in the increasing references from many sources to the fact that the Soviet Union is virtually self-sufficient in energy and materials, and this is held out as a major threat to U.S. economic and national security.[17]

The implications of the strategy are immense. First, it would involve the cutoff of the United States from its export markets and from efficient and cheap imports. This would reduce the productivity of the U.S. economy and move the economy toward a prolonged period of slow growth or stagnation. Second, it would isolate the United States politically from its allies and create very serious problems in national defense. Third, it would isolate the United States from the emerging Third World and ultimately exacerbate serious political and security problems. This is not the place to fully detail the implications of self-sufficiency and the reemergence of autarchy in the world, but it is clear that the severity of the national security issues identified here argue against the sledgehammer solution of imposed self-sufficiency. Self-sufficiency as a strategy, of course, would address both of the issues identified above.

Strategies for Supply Interruption

There are, in general, several approaches to lessen the impact of a supply interruption on those materials for which the United States depends significantly on foreign sources, including:

- maintaining stocks of the material within the continental United States;

- maintaining standby productive capacity for the material in the United States;

- developing on-the-shelf technology to substitute for the specific material in critical uses;

- fostering design changes to minimize the use of the material;

- creating substitutes for the imported material from domestic production of the same material on a subsidized basis.

Any of these strategies involves a cost to society, justified, presumably, by the benefit in mitigating the probability of a costly supply interruption. The strategy that should be followed depends then on the specific commodity situation with which the United States is faced. We have already decided in the case of materials needed for national defense that a strategic stockpile of supplies for a specific national security emergency is the best form of insurance. That strategic stockpile, however, is not useful for commercial supply interruptions and may involve substantial costs to society because it must be maintained for use in a national security crisis.

The alternative solutions to the problem of supply interruption all involve scientific and technological components. A deep-sea mining capability, for example, would immediately change the reserve and production picture for copper, nickel, cobalt and manganese and might make the United States an exporter rather than an importer of these materials. Deep-sea mining might have benefits outweighing its economic costs and therefore justify a security premium or subsidy to speed its development. Similarly, if research could yield a substitute for cobalt, based upon a more abundant material, the security premium involved in cobalt could justify our investment in the technology required to produce the substitute.

The policy issues raised by these alternatives are to a great extent specific to a given commodity and must be considered on a case-by-case basis. Much of the analysis that has been done in the critical materials area indicates that an effective and carefully drawn stockpile proposal is, in many cases, the cost-efficient insurance against supply interruptions. Other of the above policies, however, may be the cost-efficient approach in some cases. It is clear that the incentives for private research and the development of new production strategies and substitutes do not reflect society's costs in dependence upon foreign sources that can be interrupted for political purposes. Thus, there is a prima facie case that research into developing alternatives other than stockpiling indicated above should be properly undertaken at the expense of the society as a whole rather than the private sector alone.

This leads, then, to two recommendations concerning strategies for dealing with the issue of supply interruptions. The first is that for the selected materials upon which the United States is dependent on overseas sources, a strategy be established for each material to achieve protection from supply interruption in the most cost-efficient manner, given the current state of technology in production and use. Second, additional support should be undertaken for basic research in innovative technologies for continuing to produce existing materials and for creating new resource substitutes. This would be long-term strategy for decreasing the cost of insurance against supply interruption.

Strategies for the Loss of Comparative Advantage

This issue raises policy questions that run well beyond the availability of raw materials. It includes issues of productivity, industrial management, tax policy and even the rate of savings in the U.S. economy. The fundamental attack on this issue must be in research and innovation to substantially reduce the real costs of producing the primary minerals and the basic industrial products at home. This can be achieved by research aimed (1) at the production technologies themselves and (2) at producing a substitute for the basic material and manufacturing outputs at substantially lowered costs.

It has been demonstrated that, because of a basic market failure, research and development are not pursued at a socially optimal level in the United States.[18] Thus, we need to expand the level of research and development to attack the fundamental issue of the overall productivity of our industrial base. This is the third recommendation.

The Need for Better Knowledge

We are fundamentally ignorant about the geologic deposition of mineral and energy deposits in the earth's crust. The determinants of the level and the efficiency of the exploration process are not known. There is much to be learned about ways in which changes in certain institutional structures — such as nationalized firms; countries operating as entrepreneurs; and multinational, multiproduct, private corporations — affect the flow of mineral supplies. The same state of ignorance exists on the demand side of trade in resources.

The major presidential commissions that have reported on materials problems, including the draft report of the recent presidential study, all have called for increased attention to the data and analytic information available to policymakers in both the government and the private sector. The fourth recommendation of this paper is the establishment of improved data and analytic capabilities in the federal government, including substantial research in the basic geologic and social sciences directed at the mineral and energy resources sectors.

Summary

The role of depletion in international security affairs flows primarily from the relative depletion of resources in the United States; this depletion has meant substantial change in comparative advantage for both the minerals and the resources industries and for the primary industrial sectors that are based upon them. Along with geologic endowment, this situation leads to a dependence upon foreign sources for certain materials. Supply of these materials is thus subject to political interruption and to a major problem in transition from previously efficient industries to newly emerging growth sectors in the economy.

To deal with these issues in their resource and energy context, four recommendations are made:

1. Establish, for each material upon which the United States is dependent on overseas sources, a strategy for achieving protection from supply interruption in the most cost-efficient manner, given the current state of technology in production and use.

2. Undertake additional support for basic research in new production technologies for these materials and in the development of substitutes for them, as a long-term strategy for decreasing the cost of insurance against supply interruption.

3. Expand the level of basic research and development to counteract the declining productivity of our industrial base.

4. Improve the data and analytic system for materials and energy, as a guide to both the federal government and private industry.

Notes

1. W. Stanley Jevons, *The Coal Question*, 2nd ed. (London: Macmillan, 1866), preface; D. Meadows et al., *The Limits to Growth* (New York: Universe Books, 1972), introduction.

2. J. J. Shantz, *Resource Terminology: An Examination of Concepts and Recommendations for Improvements* (Palo Alto, Calif.: Electric Power Research Institute, 1975); V. McKelvey, "Mineral Resource Estimates and Public Policy," *United States Mineral Resources*, Geological Survey Professional Paper 820 (Washington, D.C.: U.S. Geological Survey, 1972).

3. H. J. Barnett and C. Morse, *Scarcity and Growth* (Baltimore: Johns Hopkins University Press, 1963).

4. V. Kerry Smith, ed., *Scarcity and Growth Reconsidered* (Baltimore: Johns Hopkins University Press, 1979).

5. E. W. Zimmerman, *Introduction to World Resources* (New York: Harper and Row, 1964).

6. T. Page, *Conservation and Economic Efficiency* (Baltimore: Johns Hopkins University Press, 1977); V. Kerry Smith, "The Evaluation of Natural Resource Adequacy: Elusive Quest or Frontier of Economic Analysis?" *Land Economics*, vol. 56, no. 3 (August 1980), pp. 257–98.

7. R. G. Ridker and W. D. Watson, *To Choose a Future* (Baltimore: Johns Hopkins University Press, 1980).

8. M. King Hubbert, "Energy Resources," in National Academy of Sciences, *Resources and Man* (San Francisco: W. H. Freeman, 1969).

9. D. A. Brobst, "Fundamental Concepts for the Analysis of Resource Availability," in Smith, *Scarcity and Growth Reconsidered*.

10. Federal Energy Administration, Office of Policy and Analysis, *Oil and Gas Resources, Reserves, and Productive Capacities* (Washington, D.C., 1974).

11. R. P. Sheldon, "Estimates of Undiscovered Petroleum Resources: A Perspective," *U.S. Geological Survey Annual Report* (Washington, D.C.: U.S. Geological Survey, 1975), pp. 11–22.

12. William A. Vogely, "Energy and Resources Planning—National and Worldwide," *Air Pollution*, 3rd ed. (New York: Academic Press, 1977), pp. 293–353.

13. Sam H. Schurr et al., *Energy in America's Future: The Choices Before Us* (Baltimore: Johns Hopkins University Press, 1979); Hans H. Landsberg and Kenneth J. Arrow et al., *Energy: The Next Twenty Years* (Cambridge, Mass.: Ballinger Publishing Co., 1979); Ridker and Watson, *To Choose a Future*.

14. Smith, "The Evaluation of Natural Resource Adequacy."

15. National Strategy Information Systems, Inc., *A White Paper: The Resource War* (Washington, D.C., 1980).

16. U.S., Executive Office of the President, Domestic Council, *Draft Report on the Issues Identified in the Nonfuel Minerals Policy Review* (Washington, D.C., August 1979).

17. National Strategy Information Systems, *A White Paper.*

18. J. E. Tilton, *U.S. Energy R and D Potion*, RFF Working Paper En-4 (Washington, D.C.: Resources for the Future, 1974).

11
Science and National Defense:
A Speculative Essay and Discussion

Kenneth E. Boulding

Introduction

This is an unusual paper that does not conform to the general pattern of the papers in this series because it deals with an almost unprecedented problem. Most of the papers deal, quite legitimately, with research priorities within an existing framework of ideas. They are within the setting, that is, of what Thomas Kuhn calls "normal science." What I am proposing is a basic parametric change in our whole view of the problem, that is, a scientific revolution. I am not asking myself, "What is the best thing to do within the existing framework?"; I am asking, "Is the existing framework adequate?" and coming out with the answer that it is not. A profound change in our whole way of thinking about national defense is necessary. Another paper could easily be written along the lines of "normal science" within the existing framework. Someone else would have to be found to do it.

This paper, like the others, was discussed among five colleagues at the AAAS Workshop on Science, Technology and International Security: Dr. Wayne Bert, Dr. Davis Bobrow, Dr. John Coleman, Dr. Richard Scribner and Dr. Lorin Stieff. I found both their written reviews and the exciting oral discussion that we had extremely helpful. What I hoped that my original paper would do was not to solve the problems but to start a discussion and to raise questions. This I feel it did among the discussants. Rather than prepare a new version of the paper, therefore, because the discussion itself is what the paper was intended to provoke, I am presenting a somewhat shortened and revised version of the original paper, in the light of some textual criticisms, without changing its essential content. Then I present a summary of the discussion and my response to it.

Kenneth E. Boulding is Distinguished Professor of Economics, Emeritus, Institute of Behavioral Science, University of Colorado, Boulder, Colo.

253

Science and National Defense

It is the business of the scientific community to perceive and to transmit into human consciousness testable images of the orderly patterns of the real world. It is the business of a science-based technology to utilize the perceptions of science to transform the real world in directions that are favorable to at least somebody's human valuations, that is, directions that in some sense are "better" rather than "worse." The orderly patterns of the real world cover systems of a wide range of complexity and structure, from the relatively simple patterns of the physical world to the greatest complexity of which we are aware, human beings and their societies.

National defense is primarily a subset of the social system insofar as it is concerned with human beings and with their technological artifacts. It also has relationships with the biological and physical sciences. Certainly the development of chemical, biological and especially nuclear weapons has had a profound impact on it. Nevertheless, it remains essentially a part of the social system; it cannot be understood except in relation to social systems. The physical and biological sciences affect the parameters of social systems but do not in themselves explain them.

Within the general pattern of the world social system, national defense deals with certain aspects of the organization and interaction of national states. National defense is a subset of the "threat system." The threat system is part of the social system in which human behavior is organized by threats rather than by exchange or by integrative structures. Because society is an ecosystem, however, all these things are related. National defense organizations, such as armed forces and departments of defense, operate in part within an exchange system in that they buy and sell things. Their capacity for survival also depends very much on the general structure of legitimacy, which is an aspect of the integrative structure. No organization can survive in a society if it is widely perceived as illegitimate, especially by those who participate in it. All national defense organizations are financed by government appropriations of money, which in turn are obtained through tax systems or through the creation of money by the state. These political structures, again, rest on what might be called legitimated threat, which again tends to break down if it is not widely felt to be legitimate.

The dynamics of legitimacy are very complex, yet also highly relevant to the problem of national defense. Legitimacy seems to come from two quite different and contradictory sources, which explains, perhaps, why it produces systems exhibiting great discontinuities. Ancient legitimacies that have persisted unchanged for a very long time sometimes collapse overnight. One important source of legitimacy is positive payoffs. Something that is perceived as clearly beneficial would tend to acquire and to retain legitimacy. The love of country, like the love of spouse, is certainly not unrelated to our perception of net benefits received from the association. Institutions that are perceived as no longer paying off and as yielding small or negative benefits are apt to lose legitimacy, as absolute monarchy did in the eighteenth century and empire did in the twentieth century.

This, however, is not enough to explain the complex dynamics of legitimacy. We also have a strange phenomenon that I call the "sacrifice trap." Negative payoffs also produce legitimacy, simply because suffering creates a sense of identity that is extremely painful to deny. If we have made sacrifices for anything, our identity becomes bound up with the objects and the purposes of those sacrifices, and it becomes extremely hard for us to admit that our sacrifices have been in vain. Unhappy marriages sometimes last longer than happy marriages. The blood of the martyrs is the seed of the church; the blood of the soldiers is the seed of the national state. Sacrifices, and the demand for sacrifice, often tend to grow until at some point they reach a discontinuity and the whole system collapses. The history of revolutions and reformations amply testifies to this.

The persistence of threat systems in human history, in spite of the fact that they probably have a very low overall payoff in terms of human welfare, has a lot to do with the curious combination of positive and negative payoffs that they involve. A threat system essentially begins with a statement on the part of the threatener to the effect, "You do something that I want, or I'll do something that you don't want." The subsequent system depends on the response of the threatened. There are at least four possible responses. The first response is submission, in which case the threat is not carried out, but the threatened party makes sacrifices and the threatener presumably benefits. Secondly, there is defiance, in which the threatened party refuses to do what the threatener wants. This moves the system back to the threatener, who then has to decide whether or not to carry out the threat. If he carries out the threat, both parties are injured. Carrying out a threat involves costs on the part of the threatener as well as damage to the threatened. If the threatener does not carry out his threat, his credibility may be impaired, so there is again the cost to the threatener. The decision obviously depends to some extent on the evaluation of these different costs when compared to the probable value of the benefits derived from the chance of submission of the threatened party.

A third possible response to threat is flight, which has been very common in human history, from the Israelites in Egypt to refugees everywhere. A fourth response is counterthreat, in which the threatened party responds by saying, "If you do something nasty to me, I'll do something nasty to you." If this results in neither threat being carried out, we have deterrence. Deterrence, however, is always subject to breakdown. It puts stress on both parties and tends to produce escalation of threat. An arms race, in which each party in the attempt to stabilize deterrence increases his threat capability, is an example. This then creates a corresponding threat increase on the part of others that may or may not reach some sort of equilibrium. Usually it does not, and systems of deterrence, while they are frequently stable in the short run, are rarely, if ever, stable in the long run. Indeed, we can argue that deterrence cannot be stable in the long run, for if it were stable it would cease to deter.

The institutions of national defense are the result of a long evolution of threat systems that is still continuing. In the neolithic, threat systems seemed to diminish in importance as opportunities for agricultural expansion increased. The rise of cities, however, in early civilization, some 3,000 B.C. or a little earlier, was clearly related to the development of organized threat systems in the shape of armies and of tax-gathering bureaucracies headed by kings, although the earliest cities seem to have been theocracies and were organized perhaps by the spiritual threats of a priesthood.

A fundamental principle of threat systems is that the size of both the area and the population that can be organized into a single system by threat is a function of the range of the instruments of threat, particularly, of course, of weapons. Obviously these are by no means the only instruments. Ancient empires depended in considerable measure on the development of mobile armies, like those of Assyria or of Alexander, which were indeed the first "guided missiles." A critical factor here is what I have elsewhere called the "loss of strength gradient."[1] The principle is that the further one is from home, the less influence one can exert. This principle expresses itself in the exchange system in terms of the cost of transporting goods and in the threat system in terms of the cost of transporting "bads." The diminution of this gradient, through a fall in the unit cost of transport, permits the development of larger organizations.

Another factor in the situation is the relationship between the development of instruments of threat, that is, capability of doing harm, and instruments of protection, which would prevent harm being done. Spears, arrows, guns and nuclear missiles are instruments of threat. Shields, armor, walls and bomb shelters are instruments of protection, which diminish the effects of the instruments of threat. Throughout human history there seems to be a constant seesaw between these two groups of instruments. The rise of technological instruments of threat expands the area of threat-based organizations; a rise in the capacity of instruments of protection can diminish it. As armies produced empires, walls produced city-states and feudal barons. Both instruments of threat and instruments of protection, however, are costly to the users, and their relative costs are very important.

One of the curious consequences of the rise of science seems to be that it has accelerated the development of instruments of threat more than the development of instruments of protection. We see evidence of this even in what might be called the eoscientific era of the late Middle Ages with the development of gunpowder, which improved the range of instruments of threat so considerably that the feudal castle was no longer a viable defense system. The feudal system collapsed and was replaced by the much larger national state, the boundaries of which could be defended, at least in the short run, by mobile armies. The European empires from the fifteenth century on were largely a result of the extraordinary cheapness of sea transport, available once a certain level of technology had been reached. In earlier times, this played an important role in developing the Roman Empire, although land transport by Roman roads also helped. The Spanish and Portuguese and later the British, French and Dutch empires were temporary products of this seapower technology. Mahan[2] pointed out that the American Revolution probably was successful because, for a brief period, the British lost command of the seas to the French.

Organized science played rather a minor role in this development, which mainly was due to improvements in what might be called "folk technology," especially in seafaring. However, science did play an important role in the improvement of maps and charts and in the development of the skills of navigation, as it had done many centuries earlier in the observations of latitude and in the eighteenth century in the solution of the problem of longitude. Even the development of steam engines and railroads does not owe very much to organized science. As has been said, thermodynamics owed a great deal to the steam engine, but the steam engine owed very little to thermodynamics. The great explosion of science-based technology began about 1860 with the development of the chemical and electrical industries, scientific metallurgy and agriculture. In the twentieth century, it continued with the nuclear industry.

This upsurge of science-based technology had an enormous impact on the technology of weaponry, especially on the range and destructiveness of weapons and their divorce from human operations. This has had a profound though confusing effect on the structure of national defense. In an age when, for the first time in human history, a unified world state has become technologically feasible, we have seen the collapse of empires and a great proliferation of independent national states, the number of which has almost trebled in the last 30 years. Nuclear deterrence has been stable now since Hiroshima and Nagasaki, for there have been no further explosions (in war) of nuclear weapons. But there is an overwhelming fear that this stability may not last, and that we are indeed sliding at an accelerating pace down a slippery slope toward a potentially irretrievable nuclear catastrophe.

The brutal truth is that a science-based technology has made the unilateral national defense of the national state ultimately a nonviable system. As long as the nuclear weapons exist, the probability of their being used is not zero. No matter how low the probability of any event, if we wait long enough, it will come off. My own highly subjective estimate is that over the last 30 years or so the probability of nuclear war has been of the same order of magnitude as that of a 100-year flood, about 1 percent or less per annum. I suspect that it rose to something like 20 percent in the Cuban crisis and is edging upward to 2, 3, 4, maybe 5 percent per annum today. These, of course, are subjective evaluations, unfortunately incapable of being tested directly and, therefore, not strictly scientific. But a lot of things that are not scientific may turn out to be true.

All this evidence suggests that we are in a very strange situation, and that we are moving toward highly unfamiliar regions of the system. The collapse of the old empires, the powerlessness of the superpowers and the many signs of the widespread erosion of the legitimacy of war as a system indicate that we may be approaching a moment of profound evolutionary change. The impotence of the superpowers was certainly seen in the case of the United States in Vietnam and in the constant frustrations of the Soviet Union in its attempts to operate in various parts of the world, such as in Egypt and now in Poland. It will be surprising if the Soviet Union does not find itself confronted with the same kind of bleeding abscess in Afghanistan that the United States found itself with in Vietnam. In Cuba, Angola and Ethiopia, the Russians find themselves with never-ending costs of support, from which it is very difficult to see that they receive the slightest benefit. As the British, the French and the Dutch found, being an imperial power does not pay. These countries have all done much better economically since they shucked off their empires. It is also likely that being a superpower does not pay, although it is taking us some time to discover this.

All this philosophy is reflected in the continual erosion of what might be called the military ethic and culture. This is reflected, for instance, in war songs. World War I produced a fine crop, World War II produced none and the Vietnam War produced nothing but antiwar songs. The tradition of military sacrifice, which goes back a very long way in human history, could be on the point of collapse. When the sacrifice involves hundreds of millions of civilians, and the military activity consists of pressing a button in a safe shelter, the end victory does not seem worthwhile.

The loss of an old legitimacy, however, can be very dangerous if it is not replaced by another; and, at the moment, certainly neither war nor peace seems to be legitimate. If the 1980 election is any indication, the U.S. people still seem to believe that a military defense will give them security, in spite of a great deal of evidence to the contrary. The scientific community has a responsibility in this matter, because it has played a highly significant role in creating the technical change in weaponry that has destroyed the unconditional viability of even the largest national state, just as much as gunpowder destroyed the viability of the feudal baron. Neither deterrence nor bomb shelters can save us in the long run. The technology of protection is fundamentally helpless in the face of the technology of destruction. A society living in bomb shelters is not worth living in, quite apart from the difficulty it would have in raising its food supply! Present-day technology offers no solution to the problem of civil defense, and only the Chinese seem to have any illusions about its feasibility. The civilian populations of the developed world are hostages to their departments of defense. They are not really defended by them. This is a condition for which the scientific community bears an inescapable responsibility because it has assisted in the production of the technology that has created it.

Does, then, the scientific community have resources within it that can answer the accusation that it has contributed to the probability of destruction of the human race as well as to its betterment? The answer to that question is perhaps a somewhat hesitant "yes." In the first place, the scientific community is a product of a very remarkable ethos, the origins of which are somewhat obscure, that attaches great value to the principle that people should be persuaded by evidence and not by threat. This renunciation of threat by the scientific community was a very important element in its remarkable success at expanding human knowledge.

The principle that the real world should speak for itself through testing and through the evidence presented by tests was something new in the experience of the human race. All previous societies relied upon threat to insure conformity of belief and practice. It is surprising how little understanding there is of this basic principle even within the scientific community, for it is practiced widely, although with occasional exceptions at the personal level. For example, in graduate schools, the unusually imaginative and creative graduate student who disagrees with the views of his professors may find himself subject to a threat system when it comes to his final examinations. This, however, is an exception and, on the whole, the renunciation of threat, particularly among peers, is perhaps the basic ethical commitment of the scientific community. If a scientist cannot persuade his peers of the truth of his views by the evidence presented, he has no other recourse.

It is to my mind a gross violation of the scientific ethic to do what the Soviet Union did in its relations with China after 1960 and what the United States is now doing in regard to its relations with the Soviet Union. Both countries are using science as part of the political threat system by withdrawing scientific contacts and communication. It is to the great credit of the Soviet scientific community that it produced a Sakharov. It is no credit at all to the U.S. scientific community that it does not seem to have produced one. Nevertheless, the scientific community has a large reserve of what might be called the "moral resource." In social systems this resource is just as important and just as real as natural resources.

The scientific community, however, has more than this. In the last 30 years or so there has developed a considerable literature and something that could properly be called a discipline in a field that is so new that even its name has not been firmly established. I like to call it "conflict studies." The French call it *polemologie*. It has gotten to the point where at least 80 colleges offer something like an interdisciplinary program in it for undergraduates. Conflict studies is still very precariously established at the graduate level, although there are a number of institutions around the world that do offer a graduate program in it. In its applied form it sometimes goes by the name of peace research or peace science, but in its purer form it transcends the political and ethical distinction between hawks and doves. In its applied form it goes well beyond the problem of international conflict and makes contributions indeed to such things as arbitration, conciliation and mediation in commercial and labor disputes and community conflict. The existence of the discipline is at least partially acknowledged in the congressional commission set up in 1980 to study the formation of a National Academy of Peace and Conflict Resolution. If such an academy comes into being, it will, of course, be a public recognition of the existence of the new discipline.

The new discipline comes out of all the older social sciences. Historically, it owes a great deal to the work of political scientist and historian Quincy Wright and to meteorologist Lewis F. Richardson.[3] The discipline that it perhaps most closely resembles is economics, for just as economics abstracts from the complexity of social life the phenomenon of exchange and related topics and inquires how exchange organizes society, conflict studies abstracts the phenomenon of conflict, which again is virtually universal in all social relationships, and studies how this organizes society. Just as economics has had an important effect on public policy — not all of it necessarily benign — over the last 200 years, from free trade to fiscal and monetary policy; so conflict studies might be expected to have substantial effect on the way conflict is conducted in order to lower the costs of conflict to all parties. Conflict processes are strongly susceptible to what might be called "perverse dynamics," that is, processes in which rational decisions on the part of each party in fact make each party worse off. The famous theoretical treatment of this is the "prisoner's dilemma" of game theory, which has received a great deal of study in recent years. To those who are familiar with it, the theory can hardly help but make a difference in the way they behave in conflict situations.

Looking now at possible contributions over the next five years, perhaps the most optimistic scenario of the present conflict environment would be like that with the Soviet Union in the Cuban crisis, in which we will move toward the cliff of nuclear war and then turn back from it. This should arouse interest in the scientific study of conflict systems, but more particularly, it should arouse interest on the applied side in the study of the management of threat systems, which is something that has been greatly neglected, even in conflict studies. First, there is a great need for careful historical analysis of threat systems of the past and the way they have been managed. There is need for a much better information system to assess the consequences of threats, and even the description of them.

One of the great problems with the threat system is that its information processes and feedbacks are extremely poor, even in comparison to the processes and feedbacks of the exchange system. In exchange we usually know fairly well what the exchange opportunities are that are open to us; thousands of prices are quoted daily in the press. The consequences of exchange are always somewhat uncertain, and exchange not infrequently results in disappointments, from which, however, we often learn rather rapidly. Once we have bought one lemon, we tend not to buy at least the same one again. In threat systems, however, the actual nature of the threats that are made is extremely uncertain, and the consequences of making them are even less certain. Threat may convey one image to the threatener and a completely different image to the threatened, of which the threatener is not aware. Under these circumstances, it is not surprising that threat systems exhibit such striking pathologies and cause an enormous amount of human misery. An improved information system with regard to threats is surely possible, even though it is by no means easy. Just as we have substantially improved the information system in economics over the last 50 years, with the development of national income statistics and indices of various kinds, we can improve the information system in threat study.

The study of weapons systems and their development has grossly neglected the place of weapons in the general threat system. The study has also overlooked the fact that a weapon is not merely a physical system but that it is also part of the social system. In physical terms, we have overkill in table knives. We certainly have enough table knives to kill everybody in the world if they fitted into a social system that demanded it, but a table knife only becomes a weapon on very rare occasions. It is at least a plausible hypothesis that we have now gotten to the point where every improvement in weaponry lessens our security and lowers our chance for survival. How to test this hypothesis without waiting for our destruction is a difficult question. However, it is not an unworthy question for scientists to ask.

As we look at possible futures, some scenarios offer hope. One is the development of general and complete disarmament through a world state. This might come about either by a strengthening of the United Nations, which seems improbable at the moment, or by what seems even more improbable – the conquest of the world by a single country. Neither of these possibilities seems very hopeful. There is, however, another alternative, which is less drastic but more realistic – the development of expanding regions of stable peace. Stable peace is a phenomenon that was virtually unknown before the nineteenth century. After 1815, however, it developed in Scandinavia, and after perhaps 1870 it developed in North America. I think we can say it has now expanded to include Western Europe and Japan. We could almost think of a broad triangle of the globe with apices at Japan, Australia and Sweden that is in the phase of stable peace. This phase of the international system involves first taking national frontiers off all agendas, except for mutually agreed adjustments. This leads to disarmed frontiers and national images that are consistent with each other. The probability of war between the constituent nations then becomes so small that it really does not enter into anybody's calculations. Stable peace is not the same as an alliance. Indeed, alliances against a common enemy do not produce stable peace, for both alliances and enmities shift. The allies of today become the enemies of tomorrow. This is not to deny that a common threat somewhere in the background may be a factor moving a group of nations into stable peace, but it is never the dominant factor.

The potentiality for stable peace unquestionably comes out of the extraordinary increase in productivity that has resulted from science-based technology. This technology has enormously diminished the comparative advantage of the threat system as a source of wealth when compared to productivity and exchange. In a technologically stagnant society like the Roman Empire, the economic gains of the threat system through conquest and plunder may have seemed attractive in the absence of any technological development. In the last 150 or 200 years, however, it has become very clear that with the effort and the cost required to extract one dollar from an exploited human being through the threat system, one could extract fifty dollars out of nature. The rise of the rich countries to wealth in the last 150 years has not primarily been the result of exploiting the poor but of the previously poor increasing their own productivity. Without this, it may well be that the conditions for stable peace might not have been developed. However, just because the underlying conditions exist does not necessarily mean that the international system itself will move toward stable peace. This requires either a set of lucky accidents, which I think was the case in North America, or, perhaps in the future, a conscious and deliberate policy directed toward producing it.

The question as to whether a research program could be set up in this area obviously depends on whether people perceive this phenomenon as belonging to the real world or whether they think that a continuation of the present system of unstable peace is inevitable and unchangeable. It is difficult indeed for people to acknowledge that an institution like unilateral national defense, which is so ancient and so well established, is in fact coming to an end because of technical change. Even if the bulk of the scientific community is not willing to acknowledge this, there still may be support for a potential alternative. A program of research along these lines, therefore, is by no means utopian. It would involve theoretical, experimental and historical research. A great deal needs to be done on the theory of threat systems. Something could be done through experimental social psychology. I confess I am personally a little skeptical about the payoffs there, but it would be worth trying. Principally, I would argue that the empirical research here has to be historical. A large-scale study of the history of threat systems, with the testing of a group of theoretical hypotheses in mind, would, it seems to me, pay off very substantially. If indeed the National Academy of Peace and Conflict Resolution is established, I would suggest that this should be its major priority.

Summary of the Discussion

Even though we did not use the term *scientific revolution*, all the discussants agreed that what I am proposing is a very basic change in the parameters of the system of national defense. They agreed it was legitimate to raise this question and that the hypothesis was at least plausible that such a parametric change was in order, in view of my contention that technological change, particularly the enormous increase in the range and destructiveness of the guided missile, had made conventional unilateral national defense unworkable in the long run.

The discussants also felt, however, and I agreed with them, that I had not dealt with the shorter-run problems, particularly with the problem of transition from the existing system into one that was ultimately more viable. Dr. Wayne Bert made the point, for instance, that people in positions of political power in almost all nations do, in fact, feel threatened by the unilateral national defense establishments. In other words, they fear the armed forces and the political apparatus for deciding to use them that exist in other countries. The only response they can think of to this threat is to set up a unilateral national defense organization of their own, perhaps as a counterthreat. Nobody actually brought up the old Roman slogan, *Si vis pacem pare bellum* (If you wish for peace, prepare for war). But it is clear that much of the motivation that creates unilateral national defense organizations and that persuades scientists and many people of good will to go into them and support them is an ineradicable fear that their nation or society will be invaded, humiliated or even destroyed by the unilateral national organizations of others, unless they have one of their own large enough to operate as a counterthreat.

The historical fact that preparing for war has very rarely, if ever, insured peace — Sweden in the last 300 years is about the only example that I can think of that is even plausible — either fails to rise into consciousness, or the fact that the price of unilateral national defense is occasional war is simply accepted as a cost that is worth the benefits derived from continued national existence and integrity. Before the development of the long-range nuclear missile, indeed, the above position was a very plausible interpretation of the condition of the world. This is not the first time in human history, however, that technical change has made a previous social orientation unviable. As I mentioned earlier in this paper, the impact of efficient cannon on the feudal system is a case in point. The scientific revolution and science-based technology destroyed both the economic and the political viability of slavery, although it took a long time to get the slave societies to recognize this. However, the question raised by Dr. Bert is a crucial one. It would involve a program of research in the field of transformation of human images of fact and value under pressure from the "real world," which at the moment we are poorly equipped to perform but which is by no means beyond the capacity of the social science community.

Dr. Davis Bobrow raised some extremely penetrating questions, which also came up in the discussion, as to how the future of national defense related to the other topics of the symposium, particularly to the problems of population, resource exhaustion and distribution of world development that were treated in other papers. There is a very important field of inquiry here. If we think of the international system as a system of "stress and strength," we can compare it to a complex network of rods, that occasionally "break" into war. If the strength of the system is greater than the stress on it, the rods do not break and there is peace. When the stress is greater than the strength, the rods break and there is war. The term *national strength* is often actually a factor in the stress that is placed on the system rather than a strength of it. An increase in national strength often increases the stress on the system.

Then the question arises, what do other large dynamic systems of population, resource exhaustion, economic development and so on do to the real strength of the world defense systems or to the stresses on them? Will differential population growth, increasing poverty and exhaustion of resources increase the stress? Do multinational corporations increase the strength of the system in the sense that they are profoundly interested in the maintenance of peace? Things like cultural exchange, scientific cooperation and even tourism may increase the strength of the system. The measurement problems here are extremely difficult. Nevertheless, the concepts are fairly clear, and one could certainly visualize a research program directed toward the impact of all the various facets of social dynamics on the overall strength and strain of the international system. This has rarely, if ever, been done.

Dr. Bobrow also suggested that useful work could be done with regard to a conceptual framework, which I had some hand in originating but which was not mentioned in the paper. This deals with the various "phase descriptions" of the international system and the circumstances that move the system from one phase into another. In two previous works[4] I outlined four possible phases of the international system. The first is "stable war," which actually is not too common in human history but is by no means unknown. The second is "unstable war," in which war is regarded as the norm but is interrupted by periods of peace, brought about by treaties, royal marriages and the like. As the intervals of peace become longer, this phase passes, often imperceptibly with no very clear boundary, into a third phase of "unstable peace," in which peace is regarded as the norm but is interrupted by periods of war. The ostensible object of this phase is to restore peace, of course on terms favorable to the victor. And then, as I argue in the paper, since 1815 at least in some areas this has passed over into areas of "stable peace," in which the probability of war is extremely low, even between independent states.

Dr. Bobrow argues that in any appraisal of research and development, particularly technological development, in weaponry, for instance, the question of whether the change may shift the system from one phase to a more adverse phase should constantly be raised. I would go even further and suggest that in a sense these phases are related to the stress-strength model. In stable war the strength of the system is virtually zero; any stress results in war. As we move from unstable war into unstable peace, the strength of the system, relative to the stress, increases. As we move into stable peace, the strength of the system becomes so great that the stress never rises to the point where the system breaks. The research problem here actually has curious parallels to the problems involved in nondestructive testing. A system break is destructive testing of the relative stress and strength of the system. When the system is broken, we know that the stress is greater than the strength. Any increase in nondestructive testing would presumably create feedback systems that would lessen the possibility of destructive testing. The information and measurement problems here are very severe, but this should be a challenge to the scientific community to vigorously pursue the enterprise rather than to abandon it.

Dr. Bobrow raises the very important short-run question of whether the development of first-strike capability in either the United States or the Soviet Union, or both, lessens the strength of the system, makes catastrophic war more probable, diminishes our security and moves us from the not-too-unstable peace that we now have into a much less stable peace. This question should certainly be an objective of any research into weapons appraisal. He also raises the question of the relationship between the ability to fight a conventional war and the probability of nuclear war. This also would seem to be a very legitimate subject of short-run research, although I am not sure myself what the outcome would be. Certainly our inability to fight a revolutionary guerrilla-type war in Vietnam only slightly increased the probability of nuclear war with the Soviet Union. Furthermore, there is a very general delegitimation of what might be called "colonial war" that makes it extremely difficult to pursue one to any politically successful conclusion, although it may take a few more examples before people are persuaded of this. An ability on the part of the United States to invade and conquer Iran would almost certainly increase the probability of nuclear war with the Soviet Union. This should certainly be a subject of research.

Dr. Bobrow's essential point, as I understand it, is that we cannot answer the question of how we get worldwide stable peace immediately, but what we can do is ask ourselves what changes in scientific knowledge and technology, and even in weapons technology, will move the system toward increasing the probability of peace and diminishing that of war. This seems to me a very sound point of view, even in the short run.

Dr. Bobrow also suggested that a paper of a very different kind could be written that would stay within the existing "normal science" framework in treating national defense. It might be that this could be done, but I could not do it.

Another important point that Dr. Bobrow makes is the necessity for studying the decision-making processes of national states with regard to unilateral national defense. I agree very strongly with this, in spite of the difficulties that are involved. He also makes the extremely interesting point that, whereas the time-horizon on technical innovations, particularly in the defense field, is apt to be from 20 to 60 years, our time-horizon in social, economic and political predictions is very much less than this. He suggests from 1 to 8 years — even 8 years seems optimistic to me. Unfortunately, I see no answer to this problem. Social systems are quite inherently unpredictable because of the fact that decisions are very frequently affected by quite random factors. They are also unpredictable because of our inherent inability to predict the future of knowledge or even of technology. If we could predict it, we would have discovered it by now. Information has to be surprising or it is not information. In social systems, information dominates the whole system and is merely modified by mechanical regularities. These considerations would suggest, however, that research and development in national defense are quite inherently pathological, that they cannot contribute to our security and must diminish it. If this is so, the urgency of finding a substitute becomes all the greater.

Dr. John Coleman made some useful specific points that I have incorporated in the revision of the original paper. He again emphasizes the point that the Department of Defense operates essentially in the short run and that its demands for research and development are largely governed by this. In reply, I would argue that I think any short-run program must be made in the light of long-run probabilities, and that if these probabilities rise to a certainty of total catastrophe, there is no point in short-run optimization. This indeed would be "sub-optimization," that is, finding the best way of doing something that should not be done at all. "Sub-optimization" is one of the major sources of bad decisions in any field of human life.

Dr. Lorin Stieff also made some excellent textual points, some of which I have tried to incorporate. He felt that the whole point of view was too unfamiliar to most people to be readily understood. He also felt the need for expanding many of the ideas in the paper, particularly the concept of the threat system and of the instability of deterrence. His criticism would suggest the need for a substantial research project on threat systems in both international and domestic society, and I would certainly endorse the need for this.

Dr. Richard Scribner shared many of Dr. Stieff's concerns and raised the question of whether any assessment of new weapons should be done by an integrated group, which would include social scientists as well as physical scientists and engineers, to look at the impact on the total world social system. He also was concerned about how to translate these rather unfamiliar ideas into language that would be comprehensible to a wider audience. He was struck with the need for the study of the legitimation of threat systems.

One problem that did not come up in the discussion was that of the economic impact of the "war industry." A good deal of work has been done on this subject over the past 25 years, and there is fairly broad consensus that this is by no means an insoluble problem. The U.S. economy especially is remarkably flexible; there is no sense in which a large war industry is necessary to produce full employment. In 1945-1946 we shifted about 30 percent of the economy from the war industry into civilian production without unemployment rising above 3 percent. In the early 1960s a fairly sharp reduction in the war industry (from about 9 percent to 7 percent of the Gross National Product [GNP]) was accompanied by an appreciable reduction in unemployment. In fact, the war industry—which has averaged about 7 percent of GNP in the past 25 years—has been a severe cumulative drain on the U.S. economy, even greater in qualitative terms than the 7 percent would suggest because of its high technology. The loss represents an internal brain-drain that probably accounts for a significant part of the relatively poor growth performance of the U.S. economy during this period.

A more difficult problem, which also did not come up, is the threat to the military subculture itself that the crisis in unilateral national defense implies. This threat is somewhat parallel to the crisis in religious subcultures that the rise of science created. Such threats are understandably sharply resisted, and creative adaptation to them is the key to survival. The potentiality for the transformation of military subcultures, such as is suggested, for instance, in Michael Harbottle's remarkable study of the United Nations' forces in Cyprus, *The Impartial Soldier*, is worthy of much serious study.[5]

In a larger context, national "defense" must be seen as part of a segment of the total world dynamic system that is concerned with the prevention of unwanted change. The social sciences in their normative mode have concentrated so much on the achievement of wanted change that they have almost totally neglected the problem of defense against unwanted change. This is unfortunate, for such a defense easily becomes pathological, as the psychological term *defense mechanisms* suggests. Nevertheless, defense in this sense is an entirely legitimate, and indeed a most important, problem in all areas of human life and interaction. The direction of the social sciences toward this kind of defense would be a most valuable widening of their agendas.

If indeed one dominant conclusion emerges from this discussion, it is that the agendas in the study of defense must be widened and that the scientific isolation of the national defense establishment must be broken down. Otherwise, we are likely to continue to slide down a slippery slope toward the cliff of irretrievable disaster in major nuclear war.

Notes

1. Kenneth E. Boulding, *Conflict and Defense* (New York: Harper, 1962).

2. Alfred Thayer Mahan, *The Major Operations of the Navies in the War of American Independence* (1913; reprinted by Greenwood Press, New York, 1969).

3. Quincy Wright, *A Study of War*, 2 vols. (Chicago: University of Chicago Press, 1942); and Lewis F. Richardson, *Arms and Insecurity* (Chicago: Quadrangle Books, 1960).

4. Boulding, *Conflict and Defense*; and idem. *Stable Peace* (Austin: University of Texas Press, 1978).

5. Michael Harbottle, *The Impartial Soldier* (London: Oxford University Press, 1970).

Appendixes

Appendix A
Workshop Participants

Workshop I: Science, Technology and International Security

Wayne Bert, International Security Affairs, Department of Defense

Davis Bobrow, Professor, Department of Government and Politics, University of Maryland

Kenneth E. Boulding, Distinguished Professor of Economics, Emeritus, Institute of Behavioral Science, University of Colorado

William C. Burrows, Agronomist, John Deere and Company Technical Center

Ronald M. Canon, Chemical Technology Division, Oak Ridge National Laboratory

John Coleman, Senior Consultant, Five Year Outlook Project, National Academy of Sciences

Ramon E. Daubon, Population and Development Policy Program, Battelle Memorial Institute

Charles S. Dennison, Executive Director, Council on Science and Technology for Development

Stanislaus J. Dundon, Congressional Science and Engineering Fellow, American Philosophical Association, Office of Representative George Brown, U.S. House of Representatives

Tomas Frejka, Senior Researcher, Population Council

James Henderson, Chairman, Division of Natural Sciences, Carver Research Laboratories, Tuskegee Institute

Christopher T. Hill, Senior Research Associate, Center for Policy Alternatives, Massachusetts Institute of Technology

Ginger P. Keller (Staff), Office of Public Sector Programs, American Association for the Advancement of Science. Current address: Graduate Program in Science, Technology and Public Policy, George Washington University

Vicki Killian (Staff), Consultant, Takoma Park, Md.

Alan Leshner, Five Year Outlook Program Manager, Office of Special Projects, National Science Foundation

Barbara Lucas, Policy Analyst, Division of Policy Research and Analysis, National Science Foundation

Melinda Meade, Associate Professor, Department of Geography, University of North Carolina

William Mills, Senior Staff Member, Office of Technology Assessment, U.S. Congress

Kathleen Newland, Senior Researcher, Worldwatch Institute

Peter Oram, Deputy Director, International Food Policy Research Institute

Richard Scribner, Science and Policy Programs Manager, American Association for the Advancement of Science

Eugene B. Skolnikoff, Director, Center for International Studies, Massachusetts Institute of Technology

Lorin R. Stieff, President, Stieff Research and Development

Conrad Taeuber, Associate Director, Center for Population Research, Georgetown University

Albert H. Teich (Project Director), Manager, Science Policy Studies, American Association for the Advancement of Science

Michael S. Teitelbaum, Program Officer, Population Office, The Ford Foundation

Ray Thornton (Chairman), President, Arkansas State University

Irene Tinker, Director, Equity Policy Center

William A. Vogely, Professor of Mineral Economics, Department of Mineral Economics, Pennsylvania State University

Jill P. Weinberg, Office of Public Sector Programs, American Association for the Advancement of Science

Charles Weiss, Jr., Science and Technology Advisor, World Bank

Sylvan H. Wittwer, Assistant Dean, College of Agriculture and Natural Resources, Michigan State University

Catherine E. Woteki, Group Leader, Food and Diet Appraisal Research, Science and Education Administration, U.S. Department of Agriculture

Christopher Wright, Science Policy Staff Member, Carnegie Institution of Washington

Workshop II: Applying Science and Technology to Public Purposes

William J. Abernathy, Professor, Harvard Business School

Willis Adcock, Assistant Vice President, Texas Instruments

R. Darryl Banks, Executive Assistant, Office of Research and Development, U.S. Environmental Protection Agency. Current address: Senior Scientist, RAND Corporation

William J. Farrell, Associate Vice President for Educational Development and Research, University of Iowa

Herbert I. Fusfeld, Director, Center for Science and Technology Policy, New York University

Denos C. Gazis, Assistant Director, Computer Science Department, IBM Research Center

Robert Gillespie, Vice Provost for Computing, University of Washington

Richard Goldstein, Department of Microbiology and Molecular Genetics, Harvard Medical School

William Hamilton, Professor, Management and Technology Program, University of Pennsylvania

Donald J. Hillman, Director, Center for Information and Computer Science, Lehigh University

Irving S. Johnson, Vice President, Lilly Research Laboratories, Eli Lilly and Company

Nathan J. Karch, Clement Associates, Inc., Scientific Regulatory Consultants

Ginger P. Keller, Office of Public Sector Programs, American Association for the Advancement of Science. Current address: Graduate Program in Science, Technology and Public Policy, George Washington University

Julia Graham Lear, Deputy Director, Community Hospital Program, School of Medicine, Georgetown University

Alan Leshner, Five Year Outlook Program Manager, Office of Special Projects, National Science Foundation

John M. Logsdon, Director, Graduate Program in Science, Technology and Public Policy, George Washington University

Leah M. Lowenstein, Associate Dean, School of Medicine, Boston University

William W. Lowrance, Senior Fellow and Director, Life Sciences and Public Policy Program, Rockefeller University

Allan C. Mazur, Professor, Social Science Program, Syracuse University

Granger Morgan, Professor, Department of Engineering and Public Policy, Carnegie-Mellon University

Pauline Newman, Director, Patent and Licensing Department, FMC Corporation

Gail Pesyna, President's Commission on a National Agenda for the Eighties. Current address: Central Research and Development Department, E.I. du Pont de Nemours & Co., Inc.

Richard A. Rettig, Senior Social Scientist, RAND Corporation. Current address: Chairman, Department of Social Sciences, Illinois Institute of Technology

Henry Riecken, Senior Program Advisor, National Library of Medicine

J. David Roessner, Professor, School of Social Sciences, Georgia Institute of Technology

Richard S. Rosenbloom, David Sarnoff Professor of Business Administration, Harvard Business School

Jane Setlow, Biology Department, Brookhaven National Laboratory

Vincent F. Simmon, Vice President for Technical Operations, Genex Corporation Laboratories

Kenneth Solomon, Engineering and Applied Science Department, RAND Corporation

Albert H. Teich (Project Director), Manager, Science Policy Studies, American Association for the Advancement of Science

Ray Thornton (Chairman), President, Arkansas State University

James W. Vaupel, Professor, Departments of Public Policy Studies and Business Administration, Duke University

Jill P. Weinberg, Office of Public Sector Programs, American Association for the Advancement of Science

Charles Weiner, Professor of History of Science and Technology, Massachusetts Institute of Technology

Karl Willenbrock, Cecil H. Green Professor of Engineering, School of Engineering and Applied Sciences, Southern Methodist University

Appendix B
AAAS Committee on Science, Engineering and Public Policy

Chairman:

The Honorable Ray Thornton (1983),* President, Arkansas State University

Dr. R. Darryl Banks (1983), Senior Scientist, RAND Corporation

Dr. Eloise E. Clark (1983)(Board representative), National Science Foundation

Dr. Gerald P. Dinneen (1983), Vice President, Science and Technology, Honeywell, Inc.

Dr. Phyllis Kahn (1984), Member, Minnesota House of Representatives

Dr. Melvin Kranzberg (1982), Callaway Professor of the History of Technology, Georgia Institute of Technology

Dr. Wesley A. Kuhrt (1982), Vice President, Technology, United Technologies Corporation

Dr. Patricia McFate (1982), Deputy Chairman, National Endowment for the Humanities

Dr. Blaine C. McKusick (1983), Haskell Laboratory for Toxicology and Industrial Medicine, E.I. du Pont de Nemours & Co.

Dr. Edwin Mansfield (1984), Department of Economics, University of Pennsylvania

Mr. Rodney W. Nichols (1983), Executive Vice President, The Rockefeller University

Dr. Gail Pesyna (1984), Program Specialist, New Business Programs, Central Research and Development Department, E.I. du Pont de Nemours & Co., Inc.

Dr. Benjamin S. P. Shen (1984), Reese W. Flower Professor of Astrophysics, University of Pennsylvania

Mr. William D. Carey (ex officio), Executive Officer, AAAS

Ms. Patricia S. Curlin, Staff Representative, AAAS, Washington, D.C.

*Terms expire on the last day of the annual meeting of the year indicated in parentheses.

Index